Full-Stack Flask and React

Learn, code, and deploy powerful web applications with Flask 2 and React 18

Olatunde Adedeji

BIRMINGHAM—MUMBAI

Full-Stack Flask and React

Copyright © 2023 Packt Publishing

Group Product Manager: Rohit Rajkumar
Publishing Product Manager: Bhavya Rao
Senior Content Development Editor: Abhishek Jadhav and Feza Shaikh
Technical Editor: Simran Ali
Copy Editor: Safis Editing
Project Coordinator: Aishwarya Mohan
Proofreader: Safis Editing
Indexer: Tejal Soni
Production Designer: Aparna Bhagat
Marketing Coordinator: Anamika Singh, Namita Velgekar, and Nivedita Pandey

First published: October 2023

Production reference: 2260923

Packt Publishing Ltd
Grosvenor House
11 St Paul's Square
Birmingham
B3 1RB, UK.

ISBN 978-1-80324-844-8

www.packtpub.com

This book is dedicated to my loving wife, Doyin, and my adorable daughters, Michelle and Mabel, with all my heart. Your unwavering support, love, and understanding have been the driving force behind my pursuit of knowledge and passion for technology.

– Olatunde Adedeji

Contributors

About the author

Olatunde Adedeji is a seasoned web developer with over 15 years of experience in developing dynamic and detail-oriented enterprise web applications. As the co-founder and principal software engineer at Mowebsite, a web and mobile applications development company, Olatunde has honed his skills in modern web development approaches to develop scalable web applications.

Along with his extensive experience in developing, maintaining, and supporting mobile, web, and enterprise applications in Python, Go, and JavaScript, Olatunde has consistently delivered excellent services as a team lead, team member, or in a consultancy capacity. He has worked across industries such as healthcare, banking and finance, information technology, education, legal, and insurance.

Olatunde is proficient in application design and solution blueprinting, including mastery of application development tools such as Flask, Django, Gin, React, and Node.js. Olatunde actively learns and researches blockchain technology and its potential to transform industries. When not weaving the web and playing chess, Olatunde spends time with his wife, Adedoyin, and his two daughters, Michelle and Mabel.

About the reviewers

Tyron Goldschmidt is a full stack software engineer. He loves working in TypeScript and Python. He has professional experience on large frontend and backed projects and currently focuses on performance optimization at Dotdash Meredith. Before turning to software engineering, he was a philosophy professor, and he has authored and edited many books for top academic publishers.

David McConkey is a full stack software engineer from Ottawa, Canada. With a background in organizational behavior, he has a detail-oriented mind and honed problem-solving skills. Equipped with excellent knowledge of JavaScript/TypeScript, React, and Flask, among many other languages and frameworks, he is able to develop tools and applications that will satisfy any user.

Acknowledgment

Before we dive into the world of *Full-Stack Flask and React*, I want to take a moment to express my heartfelt gratitude to all the amazing guys who have played a part in bringing this book to life. Without their support, expertise, and encouragement, this project wouldn't have been possible.

First, I want to thank my incredible family. Your unwavering love, patience, and understanding have been my rock throughout this entire journey. I am forever grateful for your constant encouragement and belief in me.

A big shout-out goes to the fantastic team at Packt Publishing. You folks have been phenomenal! I can't thank my editors, Mark D'Souza and Abhishek Jadhav, enough for their sharp eyes, expert guidance, and unwavering support. Your attention to detail and dedication to excellence have truly elevated the quality of this book.

To my colleagues and mentors, thank you for sharing your wisdom and experience in the world of software development. Your expertise and criticism have shaped my understanding and approach in profound ways. I also want to thank Opeyemi Opanuga. Your incredible support and words of encouragement were instrumental to the success of this book.

I also want to extend my gratitude to the countless developers and open source contributors who have dedicated their time and effort to advancing these awesome technologies, Flask and React. Your commitment to the community and passion for innovation has not only inspired this book but has also propelled the growth of modern web applications.

Lastly, I want to express my deepest gratitude to you, the reader. Your curiosity, passion, and desire to learn are what fuel authors like me to share their knowledge and experiences. I truly hope that this book equips you with the tools, insights, and confidence to embark on your own exciting adventures in full stack web application development.

To each and every one of you who has contributed in some way, big or small, thank you from the bottom of my heart. Your support and involvement have been instrumental, and I am honored to have you by my side on this incredible journey.

With sincere appreciation,

-Olatunde Adedeji

Table of Contents

Part 1 – Frontend Development with React

1

Getting Full Stack Ready with React and Flask   3

2

Getting Started with React   19

3

Managing State with React Hooks   43

4

Fetching Data with React APIs   75

5

JSX and Displaying Lists in React   87

Part 2 – Backend Development with Flask

8

9

10

Integrating the React Frontend with the Flask Backend   235

11

Fetching and Displaying Data in a React-Flask Application   249

12

Authentication and Authorization   271

13

Error Handling   291

14

Modular Architecture – Harnessing the Power of Blueprints   303

15

Flask Unit Testing   315

16

Preface

Full-Stack Flask and React is designed to be your definitive guide to mastering the art of full stack web development using the dynamic duo – React and Flask.

Whether you're a curious learner looking to learn both frontend and backend development or an experienced developer looking to expand your skill set, this book is structured to take you on a journey of building modern web applications.

In today's digital age, building web applications that are interactive, dynamic, and responsive seamlessly to user interactions is crucial. Flask, an open source, micro-web framework written in Python, promotes fast development and embraces a clean and practical design approach. On the other hand, React, a JavaScript library to build user interfaces, offers an efficient and flexible way to create interactive UIs. By combining the power of React and the lightweight Flask framework, you'll gain the ability to create modern full stack web applications with ease.

In this book, we'll head first to components-driven development with React and explore the fundamentals of React. We will explore concepts such as components, props and state, JSX, managing state with React Hooks, data fetching in React, displaying lists, events handling, routing with React Router, and unit testing with Jest. Then, we'll continue with full stack application development with Flask.

We'll cover everything from SQL and data modeling to creating RESTful APIs, integrating a React client with the Flask backend, authentication and authorization, building modular and reusable Flask applications with Blueprint, error handling, unit testing in Flask, and understanding how your web applications can be deployed to the cloud. We'll break down complex concepts into bite-sized pieces, provide clear explanations, and offer practical examples to ensure that you grasp each topic along the way. Whether you prefer learning by doing or enjoy deep dives into the inner workings of these technologies, we've got you covered.

Who this book is for

This book is ideal for developers who are interested in expanding their skills and knowledge in full stack web development. Whether you are a backend developer looking to learn frontend development with React or a frontend developer looking to learn server-side development with Flask, this book will provide you with the necessary guidance and practical examples to become proficient in full stack development.

Prior knowledge of basic HTML, CSS, JavaScript, and Python is required to benefit from this book.

Prior knowledge of basic HTML, CSS, JavaScript, and Python is required to benefit from this book.

What this book covers

This book is divided into 16 chapters, each focusing on essential concepts and practical techniques that you need to know to become proficient in full stack Flask web development with React.

Let's take a closer look at the chapters:

Chapter 1, Getting Full Stack Ready with React and Flask, lays the groundwork for your full stack development journey. We will discuss the reasons for developing web applications with React and Flask. You'll learn how to set up your development environment and install React and Flask. You will also learn the fundamentals of Git for source versioning and the project we will build in this book.

Chapter 2, Getting Started with React, introduces you to the basics of React, including exploring the React project structure, components, props, and state. We will discuss JavaScript concepts frequently used in React, such as destructuring, arrow functions, and default and named exports. You'll build a solid foundation to create React applications.

Chapter 3, Managing State with React Hooks, delves into the power of React Hooks, such as `useState`, `useEffect`, `useContext`, `useRef`, `useMemo`, `useCallback`, and `useReducer`. You'll discover how Hooks simplify state management and enable you to create more reusable and scalable code.

Chapter 4, Fetching Data with React APIs, focuses on fetching data in React applications. You'll explore different techniques using React Query, async/await syntax, the Fetch API, and Axios; handle loading and error states; and implement caching for efficient data retrieval.

Chapter 5, JSX and Displaying Lists in React, covers the display of dynamic lists in React. You'll learn how to use JSX as a bridge that connects JavaScript and HTML, implement nesting lists in JSX, loop over objects in JSX, and efficiently handle events in React.

Chapter 6, Working with React Router and Forms, guides you through the process of handling form input and validation, form submission, and implementing client-side routing using React Router.

Chapter 7, React Unit Testing, teaches you how to write comprehensive unit tests for React components using Jest and React Testing Library. You'll gain confidence in the quality and reliability of your code.

Chapter 8, SQL and Data Modeling, introduces you to SQL and database modeling. You'll learn how to set up databases, perform CRUD operations, and design efficient data models for your applications.

Chapter 9, API Development and Documentation, dives into the world of API development with Flask. You'll understand RESTful API concepts, implement CRUD operations, and document your APIs effectively.

Chapter 10, Integrating the React Frontend with the Flask Backend, focuses on establishing communication between the React client and Flask backend. You'll learn how to handle API calls and requests seamlessly.

Chapter 11, Fetching and Displaying Data in a React-Flask Application, explores the process of fetching and displaying data in a full stack React-Flask application. You'll learn how to handle CRUD operations in a Flask-React application. You will also learn how to handle pagination in a Flask application.

Chapter 12, Authentication and Authorization, covers essential topics of user authentication and authorization. You'll implement secure login and registration functionalities, identify system users and manage their information, learn about session management, create a password-protected resource, implement flash messages in Flask, and ensure the security of your application.

Chapter 13, Error Handling, delves into effective error-handling techniques in Flask applications. You'll learn about different types of errors and how to use Flask debugger, create error handlers, create custom error pages, track events in your application, and send error emails to administrators.

Chapter 14, Modular Architecture – The Power of Blueprints, introduces the concept of modular architecture and blueprints in Flask. You'll learn how to organize your code base into reusable modules and create a scalable application structure.

Chapter 15, Flask Unit Testing, explores the importance of unit testing in Flask applications. You'll discover techniques to write comprehensive tests for your Flask components, ensuring the robustness of your backend.

Chapter 16, Containerization and Flask Application Deployment, concludes the book by covering the deployment and containerization of Flask applications. You'll learn how to deploy your applications on servers and leverage containerization technologies for efficient deployment.

Each chapter is crafted to provide clear explanations to develop an interactive and efficient fast enterprise web application. You'll gain a deep understanding of both React and Flask and become proficient in developing full stack web applications.

We hope this book serves as your essential definitive guide on your journey to becoming a skilled full stack web developer. Don't hesitate to experiment, ask questions, and explore beyond the boundaries of the topics we cover. This book serves as a starting point, but the possibilities are endless. Let's dive in and unlock the limitless potential of *Full Stack Flask Web Development with React*!

To get the most out of this book

We recommend using the latest version of Python. Flask supports Python 3.8+, React, and Node.js 16+, and Docker 23+ installation is required for this book. All code and examples in this book are tested using Flask 2.3.2 and React 18.2.

Software/hardware covered in the book	Operating system requirements
Python	Windows, macOS, or Linux
React	Windows, macOS, or Linux

Software/hardware covered in the book	Operating system requirements
Flask	Windows, macOS, or Linux
PostgreSQL	Windows, macOS, or Linux
Docker	Windows, macOS, or Linux
JavaScript	Windows, macOS, or Linux

If you are using the digital version of this book, we advise you to type the code yourself or access the code from the book's GitHub repository (a link is available in the next section). Doing so will help you avoid any potential errors related to the copying and pasting of code.

Download the example code files

You can download the example code files for this book from GitHub at `https://github.com/PacktPublishing/Full-Stack-Flask-and-React`. If there's an update to the code, it will be updated in the GitHub repository.

We also have other code bundles from our rich catalog of books and videos available at `https://github.com/PacktPublishing/`. Check them out!

Conventions used

There are a number of text conventions used throughout this book.

`Code in text`: Indicates code words in text, database table names, folder names, filenames, file extensions, pathnames, dummy URLs, user input, and Twitter handles. Here is an example: "Navigate to the `Bizza` folder after the setup has finished."

A block of code is set as follows:

```
bizza/
--node_modules/
--public/
----index.html
----manifest.json
```

When we wish to draw your attention to a particular part of a code block, the relevant lines or items are set in bold:

```
function App() {
  return (
    <div className="App">
      <header className="App-header">
        <img src={logo} className="App-logo" alt="logo" />
```

```
<p>
    Edit <code>src/App.js</code> and save to reload.
```

Any command-line input or output is written as follows:

```
$ node -v
$ npm -v
```

Bold: Indicates a new term, an important word, or words that you see on screen. For instance, words in menus or dialog boxes appear in **bold**. Here is an example: "Log in and click on **Repositories**."

> **Tips or important notes**
> Appear like this.

Get in touch

Feedback from our readers is always welcome.

General feedback: If you have questions about any aspect of this book, email us at customercare@ packtpub.com and mention the book title in the subject of your message.

Errata: Although we have taken every care to ensure the accuracy of our content, mistakes do happen. If you have found a mistake in this book, we would be grateful if you would report this to us. Please visit www.packtpub.com/support/errata and fill in the form.

Piracy: If you come across any illegal copies of our works in any form on the internet, we would be grateful if you would provide us with the location address or website name. Please contact us at copyright@packt.com with a link to the material.

If you are interested in becoming an author: If there is a topic that you have expertise in and you are interested in either writing or contributing to a book, please visit authors.packtpub.com.

Share Your Thoughts

Once you've read, we'd love to hear your thoughts! Scan the QR code below to go straight to the Amazon review page for this book and share your feedback.

https://packt.link/r/1803248440

Your review is important to us and the tech community and will help us make sure we're delivering excellent quality content.

Download a free PDF copy of this book

Thanks for purchasing this book!

Do you like to read on the go but are unable to carry your print books everywhere? Is your eBook purchase not compatible with the device of your choice?

Don't worry, now with every Packt book you get a DRM-free PDF version of that book at no cost.

Read anywhere, any place, on any device. Search, copy, and paste code from your favorite technical books directly into your application.

The perks don't stop there, you can get exclusive access to discounts, newsletters, and great free content in your inbox daily

Follow these simple steps to get the benefits:

1. Scan the QR code or visit the link below

https://packt.link/free-ebook/9781803248448

2. Submit your proof of purchase
3. That's it! We'll send your free PDF and other benefits to your email directly

Part 1 –
Frontend Development
with React

Welcome to *Part 1* of our book. In this part of the book, we will embark on an exciting journey into the world of frontend development with React. In this section, we will delve into the fundamental concepts and techniques that form the backbone of modern web development using the React library.

You will learn the key principles and best practices that will empower you to build dynamic and interactive user interfaces. We will cover the core concepts, from setting up your development environment to creating reusable components and managing state in React.

This part has the following chapters:

- *Chapter 1, Getting Full Stack-Ready with React and Flask*
- *Chapter 2, Getting Started with React*
- *Chapter 3, Managing State with React Hooks*
- *Chapter 4, Fetching Data with React APIs*
- *Chapter 5, JSX and Displaying Lists in React*
- *Chapter 6, Working with React Router and Forms*
- *Chapter 7, React Unit Testing*

1

Getting Full Stack Ready with React and Flask

The creator of the first website, Sir Tim Berners-Lee, envisaged the web as an open platform that would allow internet users to share information, access opportunities, and collaborate without geographic and cultural restrictions. Interestingly, software developers are innovatively driving the realization of this mission.

As developers, we enable feature-rich web applications that make positive impacts on individuals and businesses around the world. Apart from sharing information, the web has drastically changed from mere static web pages to dynamic and database-driven web applications. Web technologists are coming up with new tools and techniques to make access to information on the internet hassle-free and natively convenient.

By the end of this chapter, you'll have a better understanding of full stack web development in the context of client-server architectures. We'll discuss major interactions that exist between the **frontend** of web applications and a database-driven **backend**.

Having these skill sets will usher you into the hall of fame of full stack web developers. This comes with complete knowledge of what it takes to start a web application development project from scratch and transform it into a full-blown web application. Whether you are a lone developer or a developer functioning in a collaborative team role, knowledge of full stack web development will boost your confidence to perform efficiently. In addition, you'll have the flexibility to fit any assigned role in a team setting.

Further, we'll dive into the reasons to use React, a UI library for building the user-facing end of web applications. You'll briefly be introduced to the world of React and the reasons why React is essential to build complex modern web application interface components that allow users to have a smooth experience.

Developing web applications requires setting up the development environments. In full stack web application development, the frontend and backend have separate development environments. We'll discuss how to set up React for the frontend and Flask as backend technology to power server-based processing and database interactions.

Additionally, we'll dive into getting ready with **Git**, which is a source version control tool that helps developers to track changes to the code base. You are expected to acquire enough basic knowledge to kickstart deploying your code to **GitHub**, an online platform for version control.

In this age of technological innovation and the proliferation of creative software developments, source version control is an integral part of development. It fosters collaboration among software developers to solve problems in open source or commercial projects.

We'll end the chapter by discussing the implementation of a real-world project we will build in this book, *Bizza*. The project takes you on a journey from a frontend web application perspective to a database-driven backend, connected to the REST API layer to facilitate communication.

So, without further ado, let's start to experience the world of full stack web application development using two in-demand tech stacks, **React** and **Flask**. By the end of this book, you will be able to develop full stack applications.

In this chapter, we will cover the following main topics:

- An introduction to full stack web development
- Why should we choose React?
- Why should we choose Flask?
- Getting ready with Git
- What are we building?

Technical requirements

The complete code for this chapter is available on GitHub at:
`https://github.com/PacktPublishing/Full-Stack-Flask-and-React/tree/main/Chapter01`

An introduction to full stack web development

Modern web applications are complex and rapidly evolving. The business community's needs and system requirements are motivating software developers to stretch beyond being able to function only as either a frontend or backend developer. The ability of web developers to develop full stack applications is now essential more than ever and on the rise.

In this book, we will focus on full stack web development, which refers to both the frontend and backend parts of web development. The frontend, sometimes referred to as the *client side*, is the visible part of any web application that users can see and interact with. The backend, sometimes referred to as the *server side*, is that portion where programmer code resides, coupled with a database and other server infrastructure.

Web developers who are skilled in both the client side (frontend development) and server side (backend development) are usually referred to as *full stack developers*.

In this book, we will use React as a library to develop an intuitive **user interface** (**UI**), or frontend, and Flask, a **microframework**, to build backend business logic components.

Let's take a closer look at these tools and the reasons we have chosen them.

Why should we choose React?

Building UIs for web applications is an essential part of web development. Interestingly, most web developers find it difficult to choose the most suitable JavaScript frontend library or framework to build UIs. In a moment, we will see why choosing React will help your career growth and projects.

React is a popular open source library, with an excellent community of developers at Meta Platforms (formerly Facebook) actively maintaining it. React is the most commonly used library according to the *Stack Overflow 2021 Developer Survey* report, in which 41.4% of professional developers stated they had used React in the past year (`https://insights.stackoverflow.com/survey/2021#section-most-popular-technologies-web-framewors`). So, what is the fuss about React?

React is simple to use to develop rich interactive interfaces for end users. You can start building reusable interface components for your web projects in no time. It is also easy to learn, as you will see in the implementation of the frontend project we'll embark upon in this book. If you are already familiar with JavaScript, learning React is really simple, as React is JavaScript-centric.

One of the main reasons React might be with us as long as the internet lives is due to its usage among technology giants such as *Facebook*, *Uber Eats*, *Skype*, and *Netflix*. Additionally, React, being a library, focuses specifically on building UI components – and nothing more. Its component-based approach to developing web and mobile applications makes it insanely popular among developers.

The further abstraction of the **Document Object Model** (**DOM**) in React to what is called the virtual DOM improves efficiency and performance in React applications. React uses special syntax, referred to as **JavaScript XML** (**JSX**), that allows you to write HTML elements in JavaScript, in contrast to the convention of putting JavaScript in HTML elements.

You don't have to learn complex templating languages such as Handlebars, EJS, and Pug. JSX helps you write React components using familiar HTML syntax with the help of a transpiler called **Babel**. Some JavaScript frameworks are very opinionated – for instance, Angular, Ember.js, and Vue. These frameworks have a rigid structural way of building web applications, unlike React, which gives you the freedom and flexibility to select your libraries, architectures, and tools.

Also, if you are interested in developing mobile applications, React Native can be a valuable tool. Your knowledge of React and its components, which seamlessly integrate with native views, empowers you to create both Android and iOS apps efficiently.

Now, let's get our hands dirty and set up environments with React.

Setting up the development environment with React

In this section, you will set up a development environment for the React application project we'll build in the course of this book.

To code and test React applications on your local machine, there are a few steps you need to take:

1. Install Node.js:

    I.   To download and install the stable version of Node.js, go to `https://nodejs.org/en/`.

    Node.js is a runtime development environment for JavaScript and, by extension, React applications. Node.js comes bundled with a command-line utility tool and package manager called **Node Package Manager** (**NPM**). Node.js and NPM are the tools required to successfully build and run any React applications.

    II. Click and download the version recommended for most users. Install by following the installation steps.

    III. To check whether Node.js was successfully installed, type the following into your command prompt:

    ```
    $    node -v
    ```

    IV. To check the version of NPM, type the following in the terminal or command prompt (`cmd`) for Windows:

    ```
    $    npm -v
    ```

The following screenshot shows that node and npm are working.

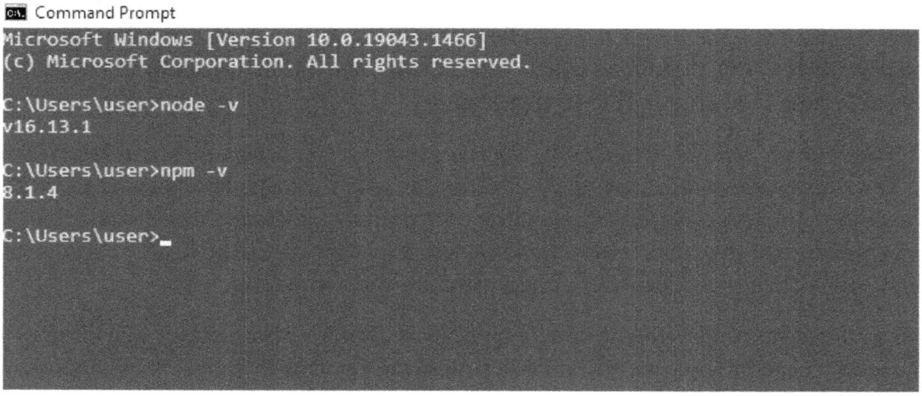

Figure 1.1 – A screenshot showing that Node.js and NPM are working

2. Install **Visual Studio Code (VS Code).**

    VS Code is a free code editor you can use to build and debug web applications. The ready-set-code approach of the VS Code editor makes it a great tool for development. VS Code has built-in support for IntelliSense code completion and features for code refactoring. Third-party extensions in VS Code with hundreds of web technologies tools allow you to be more productive and efficient.

> **Note**
> There are other code editors available to developers, but VS Code is highly recommended.

3. Install Git Client.

    Git Client is the command-line interface used to interact with Git repositories. There'll be more on Git later in the chapter. We need this tool to track changes in our project files. To install Git Client, download it from `https://git-scm.com/downloads`:

    I.   Choose your **operating system (OS)** type and install the software.

Figure 1.2 – A screenshot of the Git download page

    II. To test whether you successfully installed Git, type the following in your system's command prompt:

    ```
    $    git --version
    ```

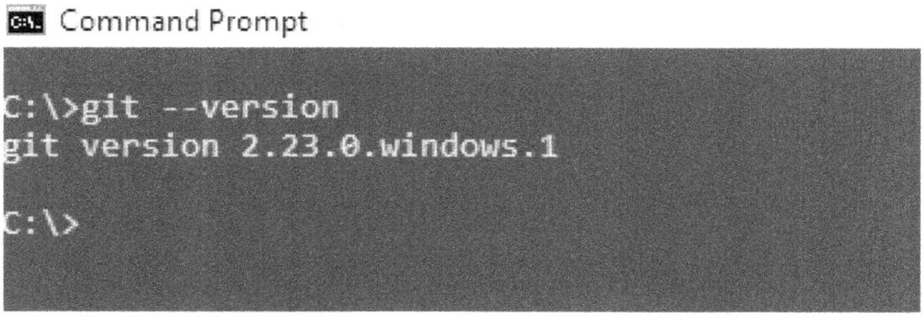

Figure 1.3 – A screenshot showing the Git Client version in Windows

We have now set up the development environment for the React applications we'll be building. This completes the frontend development environment. Let's do the same for Flask as well, and delve into why you need to choose Flask to build your backend.

Why should we choose Flask?

Flask is a minimalistic framework to develop modern Python web applications. It is a great toolkit for building enterprise-grade, scalable, reliable, and maintainable applications.

Moreover, the framework is easy to learn. Flask has no boilerplate code that must be used by developers, unlike many alternative frameworks such as Django. It is absolutely lightweight to the core. Flask as a microframework only provides developers with starting components to build web applications, while Django tends to suggest you build your web apps in a certain structure using a complete set of gears or components within its framework.

With Flask, developers have amazing freedom to choose their database, template engine, and deployment process; they can also decide how to manage users, sessions, web applications, and security.

Flask's scalability has encouraged some technology companies to migrate to Flask to efficiently implement their microservices infrastructure. A **microservice** is a small, independent, and loosely coupled software component that focuses on performing a specific function within a larger application architecture.

A microservice is like having a team of specialists, each focusing on a particular task, working together harmoniously to create something amazing. As I'm sure you would agree, cloud computing has revolutionized application development and deployment irrevocably. The science of scale that is at play with cloud computing is making it the new normal for both start-ups and enterprises. *Pinterest* is one such example.

Pinterest is one of the most visited websites in the world. It is an image-sharing and social media services platform. According to Statista, as of the fourth quarter of 2021, Pinterest had an average of 431 million monthly active users (https://www.statista.com/statistics/463353/pinterest-global-mau/). Having started their platform with the Django framework, they opted for Flask to develop their API and build a more stable microservice architecture.

Flask is still currently the main core backend technology powering Pinterest, a heavily trafficked social web application. In summary, it is easier to develop APIs and integrate varied databases in Flask. You can take it to the bank with such simplicity and flexibility assurance. If you understand Python well, then you should be able to contribute to a Flask application easily.

Flask is less opinionated, so there are fewer standards to learn. Django, conversely, gives you all you need to develop web applications – complete solutions in a box. However, the issue of scaling is what most experienced Python developers have had to deal with in their Django projects.

When you implement an out-of-the-box solution in your project, you have got a giant Django framework to deal with, which may impact negatively your project's time to market and performance.

When you combine these battle-tested technology stacks, React and Flask, in your project, you can be sure of development gains in scalability, reliability, reusability, maintainability, and secure web and mobile applications.

In this section, we discussed why you should add React and Flask to your web application's development toolkit.

Setting up the development environment with Flask

If you want to start developing web applications on your local computer with Flask as your backend framework, you need to have Python and a few other packages installed. In this section, we will set up a Flask development environment. To do so, follow these steps:

1. Install Python.

    To begin, check to see whether you have Python installed already on your computer. To test for Python installation, open Command Prompt if you use Windows or Terminal for macOS or Linux, and then type in the following command:

    ```
    $    python -version
    ```

    You will get the following output:

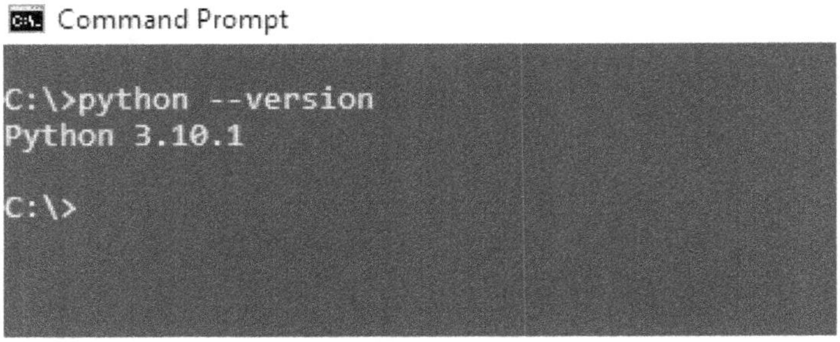

Figure 1.4 – A screenshot showing the Python version

Alternatively, you can use the following command:

```
$    python -c "import sys; print(sys.version)"
```

> **Note**
>
> In macOS or Linux, the `Python3 —version` command also works to check the Python version number and, by extension, the Python installation.

If Python has not been installed on your computer system, go to `https://www.python.org/downloads/`, choose the latest version suitable for your OS, and download and install it on your system.

2. Update `pip`.

    pip is a package manager for Python. It is a widely used tool for installing and updating packages for Python application development. If you followed step 1, you should already have pip installed along with Python:

    I.   To upgrade `pip`, type in this command on your terminal:

    ```
    $    python -m pip install --upgrade pip
    ```

    II.   Create a virtual environment.

    A **virtual environment** is a tool that allows you to have separate dependencies or package installations of your Python projects. With a virtual environment, you can conveniently have one or more Python projects on your computer system without conflicting with OS system libraries or packages from other Python installations. Every project package is self-contained or isolated in the virtual environment.

    To create a virtual environment, type the following in the terminal or `cmd` for Windows:

    ```
    $    python -m venv venv
    ```

    Alternatively, use `python3 -m venv venv` to explicitly specify Python 3 to create the virtual environment.

    For Windows users, try typing the following if you have issues:

    ```
    $    py -m venv venv
    ```

> **Note**
>
> You use `venv` for Python version 3 and `virtualenv` for Python version 2, depending on the Python version on your local machine.

3. Activate the virtual environment in Windows:

```
$    venv\Scripts\activate
```

> **Note:**
> If executing the command $ `venv\Scripts\activate` doesn't function as expected, I recommend readers to attempt using $ `venv\Scripts\activate.bat` instead.

Activate the virtual environment on macOS or Linux:

```
$    source venv/bin/activate
```

4. Install Flask:

```
$    pip install flask
```

The following screenshot shows the Flask installation command operation.

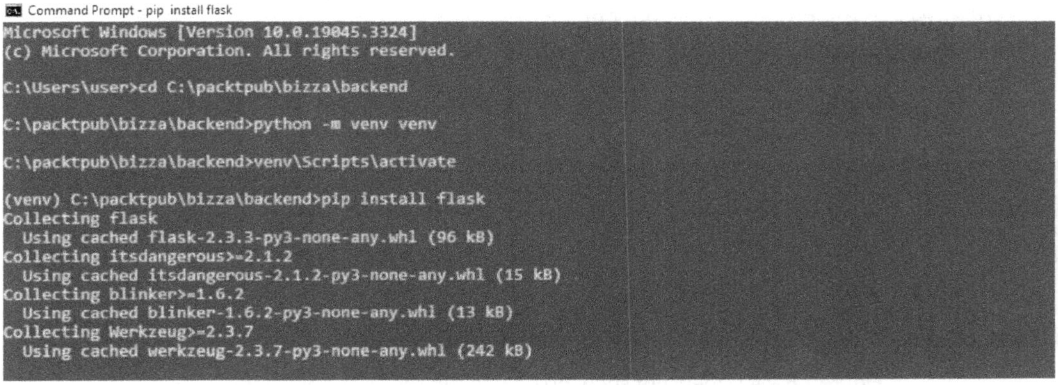

```
Command Prompt - pip  install flask
Microsoft Windows [Version 10.0.19045.3324]
(c) Microsoft Corporation. All rights reserved.

C:\Users\user>cd C:\packtpub\bizza\backend

C:\packtpub\bizza\backend>python -m venv venv

C:\packtpub\bizza\backend>venv\Scripts\activate

(venv) C:\packtpub\bizza\backend>pip install flask
Collecting flask
  Using cached flask-2.3.3-py3-none-any.whl (96 kB)
Collecting itsdangerous>=2.1.2
  Using cached itsdangerous-2.1.2-py3-none-any.whl (15 kB)
Collecting blinker>=1.6.2
  Using cached blinker-1.6.2-py3-none-any.whl (13 kB)
Collecting Werkzeug>=2.3.7
  Using cached werkzeug-2.3.7-py3-none-any.whl (242 kB)
```

Figure 1.5 – A screenshot showing the terminal commands for the flask installation

5. To test the Flask development environment, create a file named app.py in your project directory. Open the app.py file in the VS code editor and paste in the following:

```
from flask import Flask

app = Flask(__name__)
@app.route('/')
def index():
    return 'Welcome to Bizza Platform!'
```

```
if __name__ == '__main__':
    app.run()
```

6. Open the terminal and set your environment variables:

For Windows, macOS, or Linux:

I.   Create either .env or .flaskenv to store your environment variables and secrets. Inside .flaskenv, add the following:

```
FLASK_APP=app.py
FLASK_ENV=development
FLASK_DEBUG=true
```

II. Then, enter the pip install python-dotenv command in the terminal to install Python-dotenv. With python-dotenv, you can load the variables from the .env or .flaskenv file into your application's environment, making them accessible as if they were set directly as system environment variables.

III. To run the Flask app, use the following command, and you will get output similar to *Figure 1.6*:

```
$    flask run
```

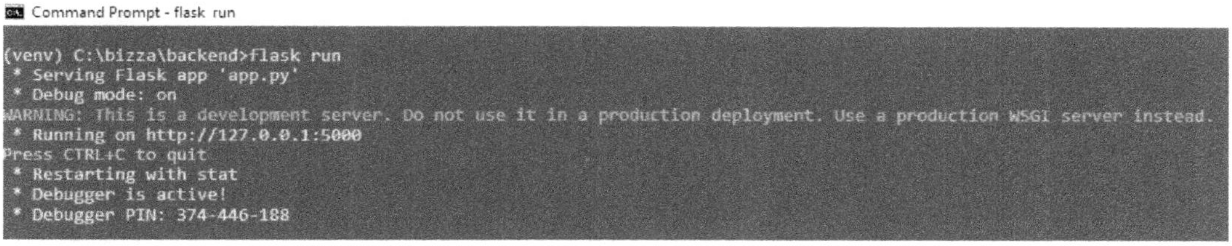

Figure 1.6 – A screenshot showing how to run the Flask application

> **Note**
> To deactivate the virtual environment, simply run $ deactivate.

Having set up and tested the development environment for Flask, we'll briefly discuss Git to understand the place of source version control in web application development and how GitHub has provided an online collaborative platform to tackle source code and encourage teamwork.

Getting ready with Git

Git is a tool for version control in software development. So, what does version control mean?

As a professional developer, you will need to write code as often as possible. Let's say you work on a project and it's 80% complete. The project lead has asked you to add a new feature to your code base, and it is urgent as the client wants your team to add that as part of the features required in the minimum viable product you will be presenting in a few days' time.

You quickly abandon what you were working on before and start working on this new feature. You make changes to one or two files to incorporate the new features. In the shortest time possible, you made the new feature work. Unfortunately, while trying to add the new feature, you unintentionally tampered with code in other files as well, and you don't even know which one of them is affected.

Now imagine you have a genie that can tell you where in your code you made your change and the exact altered line of code. Wouldn't that be awesome? Life would be super easy, saving you lots of development time. That's the magic of version control!

Version control helps you keep track of changes to the code base in your software project. It is a great way of helping developers monitor changes in their source code. Moreover, it eases the collaborative work of the development team. With version control, you can track code base changes, who is changing the code base, and when the change happens. And, if changes are not desirable, you can quickly reverse them.

Developers have used many different version control tools over the years. Git happens to be the current market leader.

How does Git work?

Git is known as **distributed version control software**. In a work environment where collaboration among team members is necessary, a complete copy of the entire source code will be available on every contributor's local computer system; we can call this a local repository.

Git tracks the local repository, maintaining a record of all the changes that occur within the local repository. It saves you the time and energy of keeping multiple versions of the project in separate local directories on your computer. This makes sharing changes to the source code between collaborators effortless.

There are three primary states in Git you should know about:

- **Modified**: In this state, files have been changed, but the changes have not yet been added to the local database by Git. These changes are the ones made since the last commit on the files.
- **Staged**: In this state, the changes have been tracked by Git and will be added to the Git local database as such in the next commit.
- **Committed**: In this state, the changed files have successfully been added to the git local database.

Let's dive deeper into version control concepts and learn how to create local and remote repositories. Before that, it will be helpful to understand the difference between Git and GitHub.

Git versus GitHub

As discussed earlier, Git is an open source tool for version control. It is simply used to track changes in a code base, track the identity of the person who made the change, and allow team coding collaboration among developers. When you set up your project on your local machine, Git is used to track changes in all the activities – adding files, updating existing files, or creating new folders.

It basically keeps a historical record of your source code. Conversely, GitHub is a cloud-based source code hosting and project management service. It simply allows you to use Git as a tool to keep your code base in a remotely hosted environment to track changes in your code base or collaboratively allow developers to work on projects.

Setting up a local repository

Install Git Client from `https://git-scm.com/download/`. Ignore this step if you already have it installed on your machine:

1. Let's start by creating a working directory named `local_repository` in the terminal:

   ```
   $      mkdir local_repository
   ```

2. Set the directory as a working directory:

   ```
   $      cd local_repository
   ```

   Inside the working directory, create an HTML file named `index.html`:

   ```
   $      touch index.html
   ```

   You will get the following output:

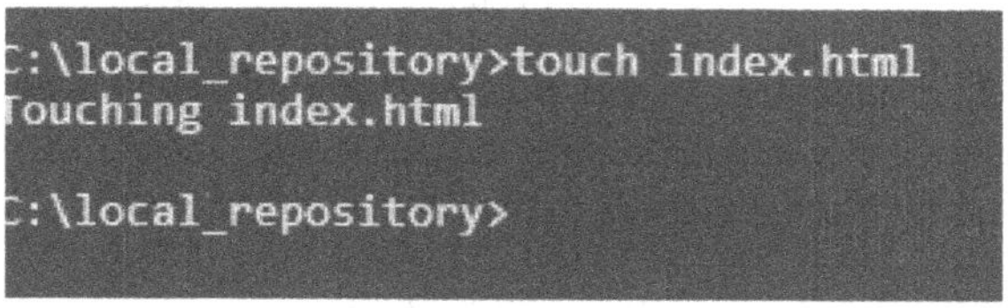

   ```
   C:\local_repository>touch index.html
   Touching index.html

   C:\local_repository>
   ```

Figure 1.7 – A screenshot creating index.html

> **Note**
>
> If you are getting an error, `'touch' is not recognized as an internal or external command, operable program or batch file`, with `touch index.html`, type `npm install -g touch-cli` in your terminal if you have Node.js installed.

3. Set up your global credentials:

```
$    git config --global user.name "Name of User"
$    git config --global user.email "test@test.com"
```

With the preceding commands, you set a global username and email address as credentials to track your contributions in the project source code.

4. Your working directory now has a new file, index.html. Enter this command in the terminal, and you will get output similar to *Figure 1.8*:

```
$    git init
```

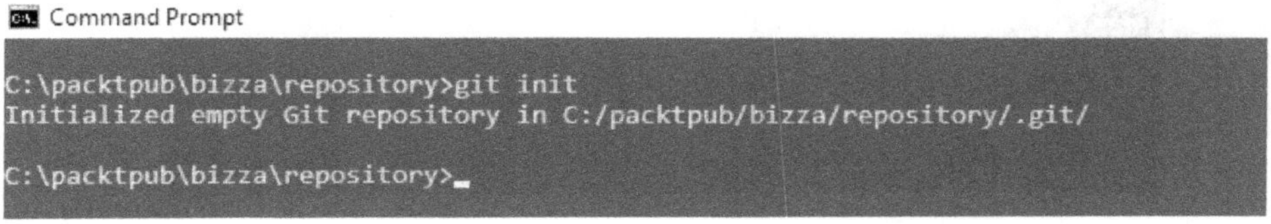

Command Prompt

```
C:\packtpub\bizza\repository>git init
Initialized empty Git repository in C:/packtpub/bizza/repository/.git/

C:\packtpub\bizza\repository>_
```

Figure 1.8 – A screenshot showing the creation of the empty Git repository

With git init, you create an empty local git repository. Git now has a local database or directory that contains all the metadata to track changes in your working directory. The .git folder is usually hidden in your working directory.

5. To add the content of your working directory to your repository, enter the following command:

```
$    git add index.html
```

This represents the staging state in Git. The changes are tracked by Git and will be added to the Git local database in the next commit.

6. To verify this, enter the following command, and you will get output similar to *Figure 1.9*:

```
$    git status
```

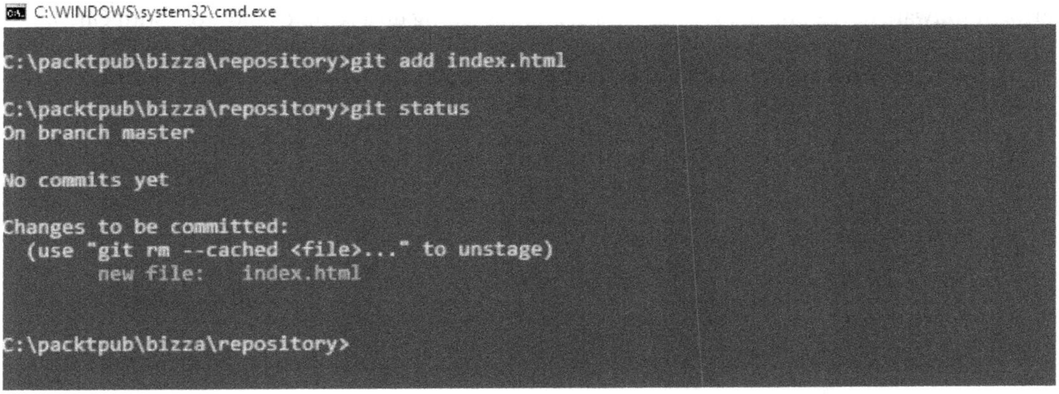

C:\WINDOWS\system32\cmd.exe

```
C:\packtpub\bizza\repository>git add index.html

C:\packtpub\bizza\repository>git status
On branch master

No commits yet

Changes to be committed:
  (use "git rm --cached <file>..." to unstage)
        new file:   index.html


C:\packtpub\bizza\repository>
```

Figure 1.9 – A screenshot showing the staging state of Git

> **Note**
> To add multiple contents, enter `git add`.

7. Now, you will need to commit to a local repository. This commit stage helps you track changes in your code base with a user-friendly message. To commit with a message flag, enter the following command, and you will get output similar to *Figure 1.10*:

    ```
    $    git commit -m "first commit"
    ```

```
C:\local_repository>git commit -m "first commit"
[master (root-commit) 95acc0b] first commit
 1 file changed, 0 insertions(+), 0 deletions(-)
 create mode 100644 index.html

C:\local_repository>
```

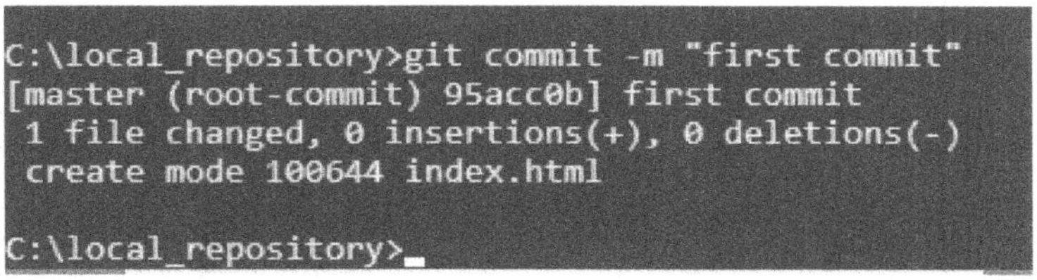

Figure 1.10 – A screenshot showing the commit state of Git

> **Note**
> It is always best practice to include a message in your commit command. This helps track changes, and if you have to revert, you can use a commit message as your saving point.

Now you understand how to create a local repository, add files to it, and transition files from the staging area to the committed state. Let's briefly discuss how you can create a remote repository on GitHub for the cloud-based storage of your source code for possible collaboration.

Setting up a remote repository with GitHub

In today's digital age, GitHub has become an essential skill for seamless collaboration and version control in software development projects. Let's delve into setting up a remote repository with GitHub:

1. Create a developer account on the GitHub site: `https://github.com/`.

https://github.com
Product
Solutions
Open Source
Pricing
Sign in
Sign up
GitHub Universe: Dive in to AI, security, and DevEx
Let's build from here
The AI-powered developer platform to build, scale, and deliver secure software.
Sign up for GitHub
Start a free enterprise trial

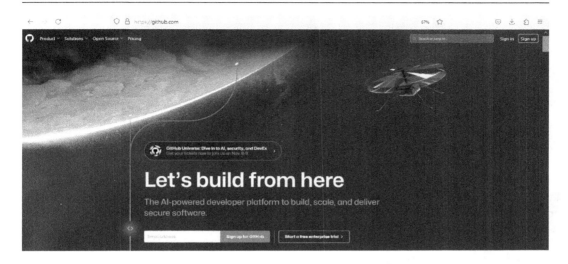

Figure 1.11 – A screenshot showing the GitHub signup page

2. Log in and click on **New**, and you will get the following screen:

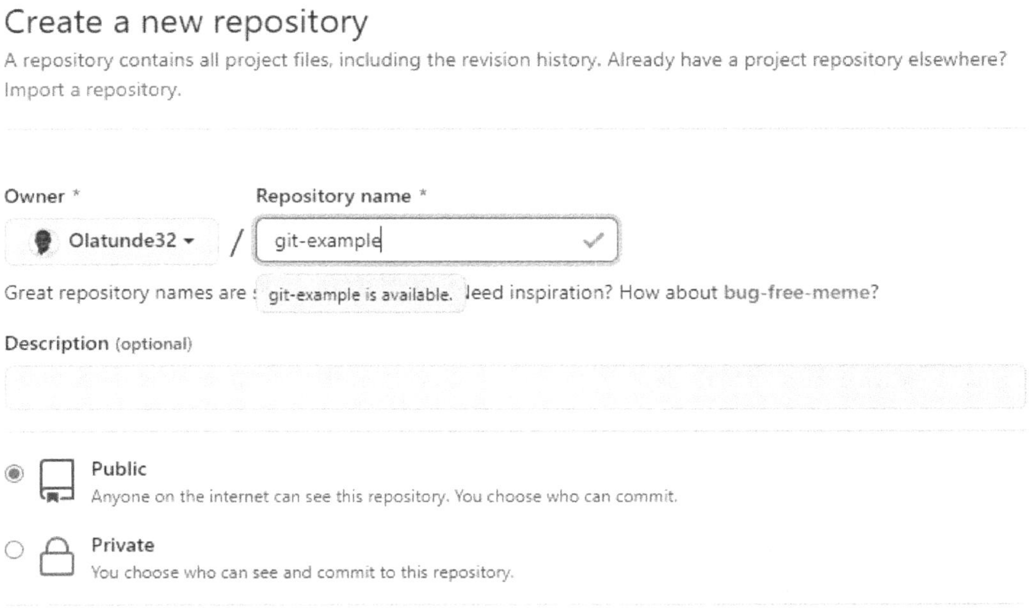

Create a new repository
A repository contains all project files, including the revision history. Already have a project repository elsewhere?
Import a repository.
Owner *
Repository name *
Olatunde32
/
git-example
Great repository names are
git-example is available.
Need inspiration? How about bug-free-meme?
Description (optional)
Public
Anyone on the internet can see this repository. You choose who can commit.
Private
You choose who can see and commit to this repository.

Figure 1.12 – A screenshot showing the staging state of Git

3.   Once you have created your new repository, enter the following command in your current working directory:

```
$    git remote add origin https://github.com/your-git-username/
your-repository-name.git
$    git branch -M main
$    git push-u origin main
```

The preceding commands move your local repository to a remote cloud-based repository to track your code base changes.

In summary, we discussed Git as a required tool for web developers in the modern age. You now know the difference between Git and GitHub. We also discussed basic, useful commands for a version control operation, either in a local or remote repository. Next, we'll dive into the real-world project we will build in this book using React and Flask.

What will we build?

In this book, we will build a full stack, database-driven web application for conference speakers. It is called *Bizza*. Users will be able to see the directory of speakers for events of interest to them, events, schedules, and the titles of the papers those speakers are presenting. The solution will include frontend development with React, authentication, authorizations, and REST API design with Flask.

We will start by implementing the frontend in the initial few chapters of the book and subsequently implement the backend in later chapters.

Summary

In this chapter, we provided a brief introduction to modern full stack web development, with an emphasis on the distinction between frontend and backend developers. We discussed the importance of React in building the user interfaces of web applications, and we explained how React and Flask are perfect tools to develop full stack web applications, owing to their simplicity, efficiency, and high performance when compared to rivals in the industry. The development environments for React and Flask were covered as well.

Lastly, we discussed the importance of Git as a tool for version control and the project we will build in this book, *Bizza*.

In the next chapter, we'll dive deeper into explaining components, props, and states in React to better understand how React applications are built. A typical React project structure will be discussed, with the aim of learning the functions of files and directories.

2

Getting Started with React

So far, we have laid a solid foundation for understanding full stack web development using modern software stacks. Hopefully, you have already set up a React development environment on your local machine. If not, you can go back to *Chapter 1, Getting Full Stack Ready with React and Flask*, and revisit the *Setting up a development environment with React* section.

In this chapter, we'll systematically start to introduce you to the world of React in a subtle way. You will learn a bunch of cool concepts and techniques that will help you develop intuitive user interfaces that enable users to interact with your web application. You will learn how to spring up your first React project without the hassle of a complex configuration and understand the basic directory structure required for every React project. Then, you'll learn how to use ES6 features in React.

Components are the building blocks of any React application. In this chapter, you will understand concepts around components and how you can use them to build user interfaces in React. This knowledge is vital for building modern frontend technologies in your project. With detailed use cases, you will understand the use of props in React for passing information between components, and how states add interactivity to React applications.

By the end of this chapter, you will have acquired, in a practical way, the skill sets required to get started with any React project. You will also have a deeper understanding of React core features (components, props, and state) to develop interactivity in any web application.

In this chapter, we're going to cover the following main topics:

- Exploring a React project directory structure
- Arrow functions in React
- Understanding what destructuring is
- Default and named exports
- What is a React component?
- What are props?
- React state

Technical requirements

The complete code for this chapter is available on GitHub at: `https://github.com/PacktPublishing/Full-Stack-Flask-and-React/tree/main/Chapter02`.

Exploring a React project directory structure

In a moment, we'll set up a React application project. There are many ways to create a React application. You could use the **Create React App** tool to generate a boilerplate application with no build configurations. This doesn't require you to have complex configuration knowledge.

You can simply focus on the implementation of your application straight up. This is what we'll be using in this book. And if you are open to exploring **Vite** (`https://vitejs.dev/`), it is another next-generation frontend tooling for rapid React application setup.

The other way of creating a React application requires knowledge of Webpack and Babel configurations. Now, without further delay, let's dive into creating our React project application. You are expected to follow along.

Open your terminal and type the following command:

```
npx create-react-app frontend
```

You will get the following output:

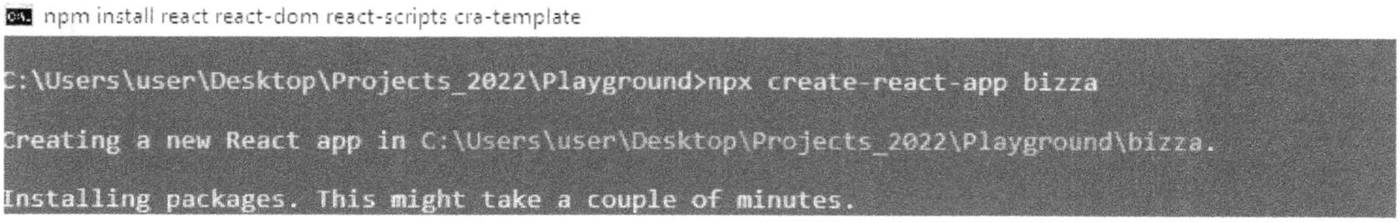

Figure 2.1 – Screenshot of the create -react-app command

Navigate into the `frontend` folder after the setup has finished:

```
$ cd frontend
```

Now, we can open the application using **Visual Studio Code** (**VS Code**) – simply type `code` . on the command line:

Figure 2.2 – Screenshot of create-react-app showing code to open
the VS Code editor in the project root directory

The following extracted folders and files structure will appear in the VS Code editor:

```
frontend/
--node_modules/
--public/
----index.html
----manifest.json
--src/
----App.css
----App.js
----App.test.js
----index.css
----index.js
--.gitignore
--package-lock.json
--package.json
--README.md
```

The following screenshot shows the screenshot of the VS Code project directory structure:

Figure 2.3 – Screenshot showing the project directory structure

So, let's quickly dive into the preceding file and folder list to understand their individual purposes:

- `node_modules/`: This folder holds all node packages installed with the Create React App tool. All the `dependencies` and `devdependencies` are stored inside. It is good to note that all subsequent packages we'll install in the future will also be kept in this folder.

- `public/`: This folder contains important public files such as `public/index.html` and `public/manifest.json`:

  - The `index` file displays on `localhost:3000` when the app is in the development environment or on a hosted domain. Essentially, this file puts React components' execution results in the `index` file's root `div` container for public view.

  - The folder also contains the `manifest.json` file, which contains application metadata and details for responsive screen display.

- `src/`: This is the most vital folder in React application development. More than 80% of your coding activities' time will be spent here. So, it is important to know exactly what you are doing inside this folder. This folder contains components and a few other files, such as `src/App.js`, `src/App.css`, `src/index.css`, `src/App.test.js`, and `src/index.js`:

  - The `src/App.js` file is used to implement React components. If you are working on a small project, you can use it to implement your application without creating other component files. All your component code will be put inside a single file. As your application grows, you might consider splitting up your components into multiple component files. With this split up, each file will then maintain one or more components.

  - The `src/App.css` file is used to style your component. The same goes for the `src/index.css` file, which is used to style your overall application. Both these files can be edited to suit your style requirements.

  - The `src/App.test.js` file is used to write unit tests for React applications.

  - The `index.js` file is the entry point to your React application.

- `.gitignore`: This file contains lists of files and folders that shouldn't be tracked and added to the Git repository in your project. For instance, `node_modules` is always listed in the file as it is only needed in the development environment.

- `package.json`: This file contains node package dependencies and other metadata used in the project. This makes it easy to set up the same on another computer system using **Node Package Manager** (**npm**) without much hassle.

- `package-lock.json`: This file stores the version number of your installed packages from npm. This ensures consistency of package installation from npm to other developers' local computer systems.

- README.md: This is a Markdown file. **Markdown** is a lightweight markup language used for creating formatted text. You can use any text editor. The README.md file helps developers to give instructions and essential information about their projects. On GitHub, you can use the README.md file to display information about the content contained in project repositories. The Create React App tool generates this file for us out of the box.

Now that you know some of the purposes of the folders and files structure of your React project, let's run the npm start command in the terminal to see the default React application:

Figure 2.4 – Screenshot of the React app home page on localhost:3000

> **Note**
>
> The following shows some of the messages such as success messages, available scripts, and how to start the development server after the installation of React:
>
> - npm start: Starts the development server
> - npm run build: Bundles the app into static files for production
> - npm test: Starts the test runner

In summary, we have discussed how you can power up your React application with the create-react-app tool. You now know what each folder and file represents in detail. With this, you can be confident that you know how to set up a React application. Let's start the next section by discussing arrow functions. Arrow functions are one of the features of **ECMAScript 2015 (ES6)**. They're a nifty addition to JavaScript that makes writing functions a breeze!

Arrow functions in React

Arrow functions provide a more concise and readable syntax for defining functions in JavaScript. Arrow functions have become a widely used feature in React development for obvious reasons: shorter syntax and implicit return. You will have a better understanding of what these mean in a moment.

In traditional JavaScript, you would have to define a regular function that adds two numbers together like this:

```
function addNumbers(a, b) {
    return a + b;
}
```

Not bad, right? But arrow functions can make this even simpler and more elegant. Check this out:

```
const addNumbers = (a, b) => {
    return a + b;
};
```

Pretty cool? The `function` keyword is replaced with a funky-looking arrow, `=>`, and you can omit the curly braces and the `return` statement if your function is just a one-liner:

```
const addNumbers = (a, b) => a + b;
```

Basically, you define arrow functions by following this syntax rule:

```
const functionName = (parameters) => {
    return returnValue;
};
```

Or, for even shorter function definition with implicit return when the function body consists of a single expression, use this rule:

```
const functionName = (parameters) => returnValue;
```

Arrow functions are often used within React applications due to their concise syntax and benefits, especially when defining function components, handling event callbacks, and mapping arrays. We will see more arrow functions usage in the course of this book.

However, it's important to note that arrow functions are not a complete replacement for regular functions. They have some limitations, such as not having their own `this` context, which makes them unsuitable for certain use cases, such as defining object methods.

When used in defining the object method, arrow functions do not have their own context and would not properly access the name property within the object. Let's use the following code example to illustrate it better.

```
const speaker = {
    name: 'Alice Andrew',
    sayHi: function () {
        console.log(`Hi, I'm ${this.name}!`);
    },
};
speaker.sayHi(); // Output: Hi, I'm Alice Andrew!
```

Now, let's use an arrow function to define the object method:

```
const speaker = {
    name: 'Alice Andrew',
    sayHi: () => {
        console.log(`Hi, I'm ${this.name}!`);
    },
};

speaker.sayHi(); // Output: Hi, I'm undefined!
```

In summary, arrow functions are a fantastic feature in ES6, making your code cleaner and more pleasant to work with. They're perfect for short functions and can make your life as a React developer a whole lot easier.

Next, we will take a look at another cool ES6 feature: **destructuring**. You will need the destructuring technique to be able to extract values from arrays or objects in a more concise and readable way.

Understanding what destructuring is

Destructuring is simply a JavaScript expression that ensures you are able to extract multiple values from arrays or properties from objects into separate unique variables. Destructuring is one of the awesome JavaScript ES6 features used in React to manage data.

ES6 was a major milestone in the effort to improve standards in the JavaScript language. Destructuring takes extracting data from arrays and objects to a new concise level. Prior to ES6, you could declare and extract data from an array this way:

```
const speakers = ["John", "Walke", "Dan", "Sophie"];
const firstspeakerName = speakers[0];
const secondspeakerName = speakers[1];
const thirdspeakerName = speakers[2];
```

```
const fourthspeakerName = speakers[3];

console.log(firstspeakerName); // "John"
console.log(secondspeakerName); // "Walke"
console.log(thirdspeakerName); // "Dan"
console.log(fourthspeakerName); // "Sophie"
```

With destructuring, the code looks like this:

```
const speakers = ["John", "Walke", "Dan", "Sophie"];
const [firstspeakerName, secondspeakerName,
    thirdspeakerName,fourthspeakerName] = speakers
console.log(firstspeakerName) // "John"
console.log(secondspeakerName) // "Walke"
console.log(thirdspeakerName) // "Dan"
console.log(fourthspeakerName) // "Sophie"
```

If we want to skip "John" in the array and write out to the console the remaining items in the array, we can do the following:

```
const speakers = ["John", "Walke", "Dan", "Sophie"];
const [, ...rest] = speakers // the … used is called the
                              spread operator
console.log(rest)// output: "Walke", "Dan", "Sophie" John
                   will be skipped
```

In addition, it is not only arrays that can be destructured; you can perform destructuring in objects as well. For instance, the following code shows how to destructure object properties:

```
const speakers = {
    id: 1,
    name: "Juliet Runolf",
    jobTitle: "Director, Marketing",
    company: "Abernatny Group",

    address: {
    street: "Okaland Dtuse",
    city: "Greenland",
    state: "Houston",
    country: "USA",
    }
}
function App()
{
    const {name, jobTitle, company} = speakers;
```

```
//destructuring of object properties
return (
    <>
        <div>
            <h2>Name: {name}</h2>
            <h4>Position: {jobTitle}</h4>
            <h4>Company: {company}</h4>
        </div>
    </>
);
}
```

You can see how we were able to extract the values of properties in the preceding code snippet. It is even possible to destructure the nested address property in the object:

```
const {name, jobTitle, company, address} = speakers;
// destructuring of object properties
const {street, city, state, country} = address;
// destructuring of nested address property
    return (
    <div> <h2>Name: {name}</h2>
        <h4>Position: {jobTitle}</h4>
        <h4>Company: {company}</h4>
        <h4>Street: {street}</h4>
        <h4>City: {city}</h4>
        <h4>State: {state}</h4>
        <h4>Country: {country}</h4>
    </div>
    );
}
```

So, what are the benefits of destructuring in React applications? Destructuring makes your code more compact and easier to understand. It allows you to extract specific properties from objects or elements from arrays directly, reducing the need for repetitive dot notation or array indexing.

Also, destructuring allows you to set default values for properties – for instance, take a look at the following line of code:

```
const { name = 'Anonymous', age = 0} = speaker
```

The preceding code can be useful when working with optional or potentially undefined data and default values are required. Destructuring is used in accessing props and state in components. Oh, you don't know what components, props, and state are all about? Fret not, we will discuss these concepts in this chapter shortly.

In summary, destructuring allows programmers to access data, even complex nested data in arrays and objects, in an easy way. It substantially improves the quality of code readability and access. And if you want a shorter line of code in your React application, destructuring clearly helps. It helps to cut the amount of code used in an application.

The next cool concepts to understand in React application development are default and named exports. Let's dive in and understand these React concepts.

Default and named exports

As mentioned earlier, ECMAScript 2015, also known as ES6, was a major milestone in the effort to improve standards in the JavaScript language. Among the new features added were modules and the ability to use `import` expressions. Modules allow us to better organize our code base into logical units. Basically, modules could be a function or related functions designed to perform specific tasks. They make code reusability across projects easier.

In React, we use a default export to make component functions, variables, classes, or objects available to other component files. Only one default export is allowed per file.

For instance, the following code makes it possible to import a file from the `Speaker` component:

```
import Speaker from './Speaker';
```

The following code makes it possible to export the file to another component file:

```
function App()
{
    return (
    <div>  …   </div>
    );
}


    export default App; // Specifically, this code line
                        makes it possible to export the
                        file to another component file
```

In a named export, you can have multiple named exports per file. And when you want to import, you can name the specific import with braces, like this:

```
import { FirstComponent, SecondComponent } from "./ThirdComponent";
```

To sum up, default and named components are a way to make specific functions available across components in any React project.

Next, we'll delve into the core essence of React components and gain a clear understanding of their purpose and functionality.

What is a React component?

A **component** is the core building block of any React application. Sometimes, you could perceive React as a JavaScript coated with some chocolate. Chocolate is sweet I guess, and so is React. Seriously, building a UI with vanilla JavaScript can be cumbersome. You can struggle with the expensive DOM dilemma!

The thing is, when working with vanilla JavaScript to handle the **Document Object Model (DOM)**, it can get pretty expensive – both in terms of time and effort. Frequent DOM manipulation is high in non-React applications, which ultimately results in the slowness of website elements update.

The **virtual DOM** solved this problem in React. The DOM only updates what was changed, not the entire DOM tree. However, if you remember how you use functions in plain JavaScript, writing components won't be a challenge. A function in JavaScript is a code block designed essentially to perform certain tasks.

The same applies to React components, which are reusable, maintainable, and self-contained blocks of code to return a UI. In React, components return HTML elements mixed with some JavaScript.

There are two types of React components: **class components** and **function components**. In this book, we'll adopt function component coding patterns to develop our React application.

Function components are the present and future of React, so if you are just learning React, the function component is easier to learn without the extras involved in writing class components. And if you are already familiar with class components in React, you can still use function components along with class components. If you find yourself working with a legacy React code base, it's definitely worth considering a gradual migration toward function components.

Let us learn how to define a function component and use it in React applications by doing the following steps:

1. Open `App.js` in the `src/` file of the React project directory. The `src/App.js` file contains the following code:

```
import logo from './logo.svg';
import './App.css';

function App() {
    return (
        <div className="App">
            <header className="App-header">
                <img src={logo} className="App-logo"
                    alt="logo" />
                <p>
                    Edit <code>src/App.js</code> and
                        save to reload.
                </p>
```

```
            <a
                className="App-link"
                href="https://reactjs.org"
                target="_blank"
                rel="noopener noreferrer">
                Learn React
            </a>
        </header>
    </div>
    );
}

export default App;
```

2. Let's remove all the boilerplate code in the file to easily understand the code structure. Replace the preceding code snippet with the following code snippet:

```
function App() {
    return (
        <div>
            <h1>Welcome to Bizza Platform</h1>
        </div>
    );
}

export default App;
```

> **Note**
> Your `src/App.js` file should now look like the preceding snippet.

3. Save the file, start your application with `npm start` in the command line, and you will get the following output:

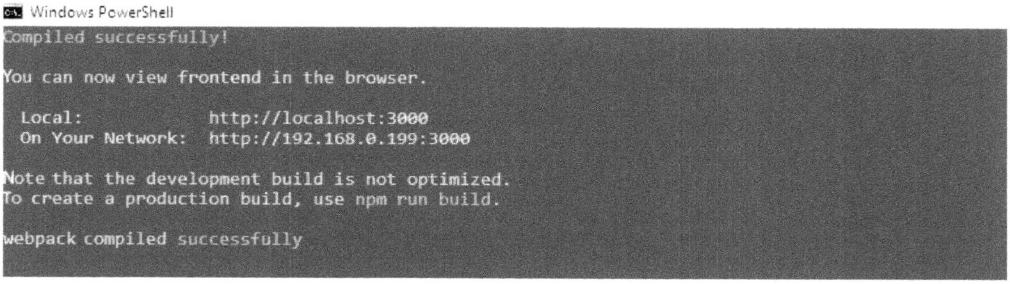

```
 Windows PowerShell
Compiled successfully!

You can now view frontend in the browser.

  Local:            http://localhost:3000
  On Your Network:  http://192.168.0.199:3000

Note that the development build is not optimized.
To create a production build, use npm run build.

webpack compiled successfully
```

Figure 2.5 – Screenshot showing the output of the npm start command

4. Check whether **Welcome to Bizza Platform** displays on your browser. If yes, you are still on the right track.

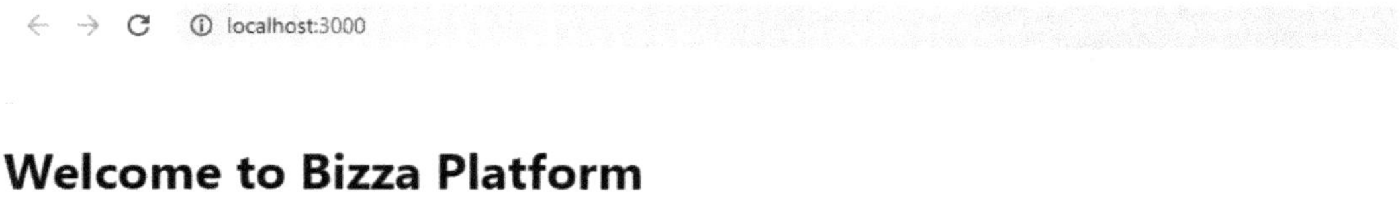

← → C   ⓘ localhost:3000

Welcome to Bizza Platform

Figure 2.6 – Screenshot of the React app home page on localhost:3000

So, let's dive deeper into each element of the code block to understand function components better.

Function components

As stated already, function components are conceptually a typical JavaScript function with the ability to receive data as props and return HTML elements in the form of JavaScript XML.

In the preceding code, the App component doesn't have any parameters yet in its function definition.

This is how you define an App component in a React application:

```
function App() {.....}
```

The code returns the following HTML elements:

```
return (
    <div>
        <h1>....</h1>
    </div>
);
```

The App component returns HTML code. The returned HTML code is a mixture of HTML and JavaScript. This is called **JavaScript XML (JSX)**. JSX is a syntax extension used in React that allows you to write HTML-like code directly within JavaScript. JSX makes it easier to describe the structure of user interfaces in React components. In *Chapter 5, JSX and Displaying Lists in React*, we'll discuss JSX in depth.

As discussed earlier, we'll be focusing more on function components in this book. If you are comfortable writing functions in plain JavaScript, function components in React will definitely be familiar. In React function components, you are able to have implementation details between the function definition and the `return` statement.

For instance, let's check App component inside `src/App.js`:

```
function App() {
    // you can perform some operations here.
```

```
    return (
        <div>
            <h1>Welcome to Bizza Platform</h1>
        </div>
    );
}
export default App;
```

Any variables defined in the function's body will be re-defined each time this function runs:

```
function App() {
    const speakerName = 'John Wilson';
    // variable declared inside function component body
    return (
        <div>
            <h1>Welcome to Bizza Platform, {speakerName}
                </h1>
        </div>
    );
}
export default App;
```

Start your browser with npm start and check the updated content in your browser.

If your server is still running, you do not need to use npm start again. Your app will recompile and display an updated view as soon as you save your file:

← → C ⓘ localhost:3000

Welcome to Bizza Platform, John Wilson

Figure 2.7 – Screenshot showing the preceding snippet output

Also, you can have your variable defined outside the body of your component:

```
const speakerName = 'John  Wilson';
// variable declared outside function component body
function App() {
    return (
        <div>
            <h1>Welcome to Bizza Platform, {speakerName}
                </h1>
        </div>
```

```
    );
}
export default App;
```

In the function component, the results from the function operation can either be *run the first time* or *re-rendered on updates*. However, you can define your variable outside the function component if it doesn't need anything from within the function component's body; otherwise, consider defining it in the function component's body.

In summary, we have been able to discuss function components in React and how to write and display the content of your function component. Next, we will discuss class components and why they are called **stateful components**.

Class components

React provides us with the flexibility to construct UI components using either functions or classes for the component code. Class components are JavaScript classes that extend React.Component and call a render method that returns an HTML element. Classes are stateful and were the only way to manage state in React before the React team came up with Hooks. More on Hooks later, in *Chapter 3, Managing State with React Hooks*. Let's look at the class code syntax for writing minimal React class components.

Create a SpeakerProfile.js file in src/ and type in the following code in src/ SpeakerProfile.js:

```
import React from 'react';
class SpeakerProfile extends React.Component {
    render() {
        return <h1>This is the class component expression
            from Speaker Profile!</h1>;
    }
}

export default SpeakerProfile;
```

Let's dive into each line of code to understand what they do:

- import React from 'react': This allows us to use the core functions of React in our code file
- class SpeakerProfile extends React.Component { }: This allows us to create the SpeakerProfile class component that inherits from the React.Component base class
- render() {   return <h1>...</h1>;   }: Every class component must have a render function that returns an HTML element

The preceding code lines explain the basic structure of React class components.

In `src/App.js`, add the following:

```
import React from 'react';
import SpeakerProfile from './SpeakerProfile';

function App(){
    return (
        <div style={{backgroundColor: 'gray', margin:20,
            color:'white'}}>
        <SpeakerProfile />
        </div>
    );
}
export default App;
```

If your `http://localhost:3000/` server is still running, you will see the following screen. If not, open your terminal and go to your working directory and type the `npm start` command:

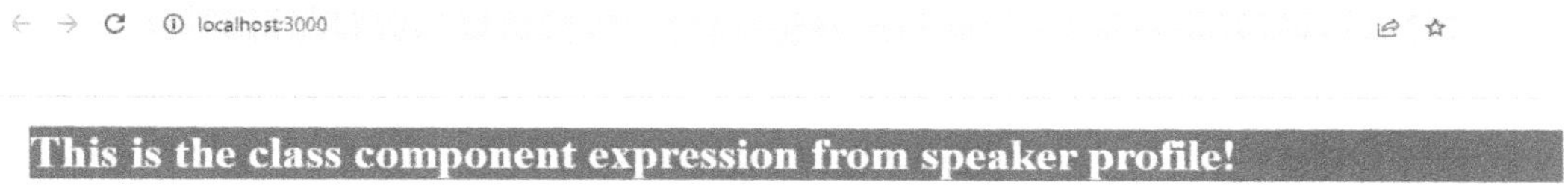

This is the class component expression from speaker profile!

Figure 2.8 – Screenshot showing output for class component

> **Note**
>
> The `<SpeakerProfile />` component was added to `src/App.js` for rendering the content of the `SpeakerProfile` component.

Let's compare the function component equivalent to the class component we wrote in `src/SpeakerProfile.js`:

```
import React from 'react';
const SpeakerProfile =()=> {
return <h1>This is a function component equivalent to a
    class component !</h1>;
}
export default SpeakerProfile;
```

So, you can see that the function components take fewer lines of code – meaning they can be elegant, concise, and readable. When using function components, you don't need to define a `render()` method, and the component itself is the function that returns JSX, making the code simpler and easier to read.

Let's briefly discuss the subtle differences in class and function components.

Class component versus function component

When creating a class component, you must inherit it from `React.Component` with the `extend` keyword and create a `render` method that is responsible for returning a React element. In the function component, there is no `render` function. It accepts `props` as an argument and returns HTML elements.

Class components are stateful when compared to function components, which are stateless. The only way you could make a function component stateful was to rewrite the code as a class component before Hooks came to its rescue with React version 16.8. Now, with Hooks, function components can be stateful.

In class components, you can utilize React life cycle methods such as `componentDidMount()`, `componentDidUpdate()`, and `componentWillUnmount()`. However, in React function components, you cannot use these life cycle methods.

In summary, we discussed class components in React and how to write and display the content of your class component. Next, we will be discussing the component life cycle to better design reusable and maintainable React components.

Component life cycle

So far, we have seen the component-oriented nature of every React application. Components interact with components to provide an interface for user interaction in web applications. However, a component – the building block of every UI we interact with – has a life cycle.

As in life, there are different phases involved; we are born, then we grow, and then we die. React components experience phases as well. They undergo three phases: mounting, updating, and unmounting. Each of these phases will be discussed briefly.

Mounting phase

This represents the birth phase of a React component. This is the phase when a component instance is created and inserted into the DOM. There are four methods present in this phase, which are listed as follows in their order:

1. `constructor()`: This method is called during the mounting phase and actually before the component is mounted. Calling the `constructor()` method accepts `props` as arguments. This is followed by calling `super(props)` as well. The `constructor()` method, however, serves two main purposes:

    -   Initializes a local state in the class component by assigning an object to `this.state`

    -   Binds event handler methods

2. `static getDerivedStateFromProps()`: This method is invoked immediately before rendering the elements in the DOM.

3. `render()`: This method is the most vital and it is always required for outputting React HTML elements. It normally injects HTML into the DOM. What you see on your interface depends on what this `render` method returns in the form of JSX.

4. `componentDidMount()`: This method is initiated after the React component is rendered into the DOM tree, essentially after the first `render()` invocation. The `componentDidMount()` method allows you to execute user actions and other side effects once components are mounted. For instance, the `componentDidMount()` method can run statements that require the loading of data from external sources. Also, you can trigger a user or system event when components had been rendered.

Updating phase

This is the hypothetical growing phase of the React component. This phase occurs immediately after the mounting phase of the component life cycle. The `componentDidUpdate()` method is commonly used in this phase. A React component is updated whenever there is a change in data state or props. The state is re-rendered using the `render()` method, and the updated state is returned to the UI.

Unmounting phase

This phase is regarded as the death phase of the component life cycle process. The `componentWillUnmount()` method is invoked in this phase immediately before the component is unmounted from the DOM tree.

Typically, web applications have many user interfaces, such as buttons, form input, accordions tabs, and even a navigation bar in our application. Users tend to move from one component interaction to another and, on the whole, have a reactive experience.

So, when users stop interacting with a component – say by moving on from a *contact* component to an *accordion tab* on the home page of our application – that switch from the *contact* page to an *accordion tab* on the home page spells the death of the life cycle to the *contact* page component.

At every point in the user interaction cycle, the React components are either inserted into the DOM tree or state changes, or the completion of component's lifecycle.

The following figure shows the React life cycle methods:

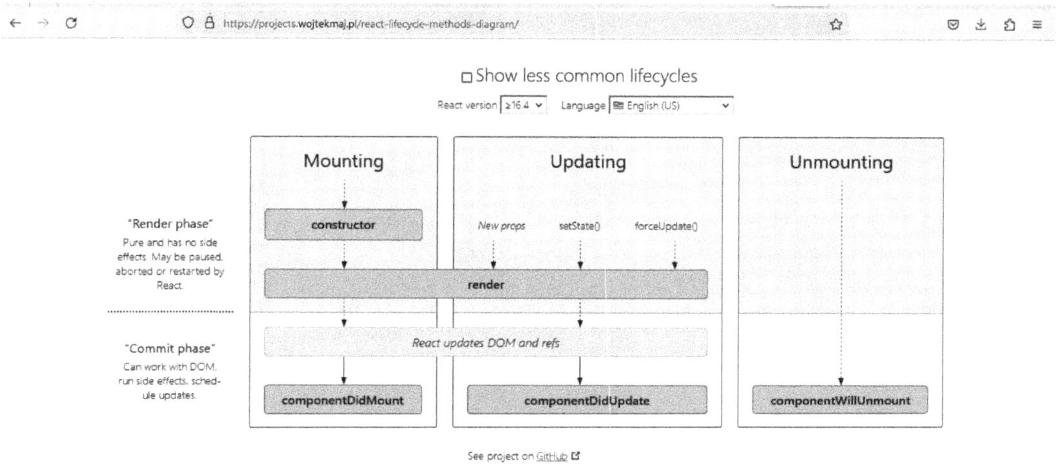

Figure 2.9 – React life cycle methods (source: https://projects.
wojtekmaj.pl/react-lifecycle-methods-diagram/)

In summary, we have briefly discussed various life cycle methods in React class components and why each of them exists in rendering React components. Later in the book, we'll explore how the life cycle is handled in a function component. Meanwhile, we will be discussing **properties** (**props**) in React. Props are one of the key features that allow React components to communicate with other components, making React applications so reusable.

What are props?

Props are arguments passed to React functions and class components. If that is too techy, let's break it down a bit. Basically, you use props to pass data from one component to another component. So, props refer to objects that store the value of attributes. This is similar to HTML when you pass a value to an attribute of a tag.

In the React world, props are used to customize and configure components and are passed from parent to child down the component tree. This means the parent component can only pass information to the child component. This is a unidirectional data flow concept in React. In essence, props are read-only, meaning that the component receiving them cannot modify their values directly.

Passing data as props

Let's take a look at an example where props are used in React components. As we have discussed, props are used in React to pass information from one component to another. In the following snippet, we will explain how data is passed as props in React.

In `src/App.js`, add the following:

```
import React from 'react';

function SpeakerProfile(props) {
    return(
        <>
            <h3>{props.name}</h3>
            <p>Position: {props.jobTitle}</p>
            <p>Company: {props.company}</p>
        </>
    );
}
//Parent component
function App() {
    return (
        <>
            <h1>Speaker Profile</h1>
            // Child component with attributes
               name,jobTitle and company inside parent
               component
            <SpeakerProfile
                name='Juliet Runolf'
                jobTitle='Director,
                Marketing' company='Abernathy Group'
            />
        </>

    );
}

export default App;
```

In the preceding code, we added a child component, `<SpeakerProfile />`, into the parent function component, `App.js`, and have some props passed to the `<SpeakerProfile/>` component.

Now, let's dive into the internal structure of the `SpeakerProfile` snippet:

```
const SpeakerProfile = (props) =>{
    return(
        <>
            <h3>{props.name}</h3>
            <p>Position: {props.jobTitle}</p>
            <p>Company: {props.company}</p>
        </>
```

```
  );
}
```

`SpeakerProfile` is a child component. We define a `SpeakerProfile` function and pass data props to it in the parent component. In the `SpeakerProfile` component body, we return HTML elements, `<h3>{props.name}</h3>`.

We then pass attribute values to these properties (props) from the `App` parent component to the `SpeakerProfile` child component:

```
<p>Position: {props.jobTitle}</p>
<p>Company: {props.company}</p>
```

The following shows the screen output for the props passed to the `SpeakerProfile` component:

Speaker Profile

Juliet Runolf

Position: Director, Marketing

Company: Abernathy Group

Figure 2.10 – Screenshot showing props in a component

In summary, we learned how to pass information from one component to another component in a function component. The flow of information from the parent component to the child component is expected to be unidirectional as a rule in React: from parent to child component.

Props are read-only. Child components cannot mutate directly. Later in the chapter, we'll discuss how this rule can be overridden by passing props from the child to the parent component.

Now, we will discuss state in React as a way to make React components dynamic and interactive.

React state

State is a built-in object in React that is used to hold information about components. It is what is responsible for the interactivity of components. In the React application, state changes. When there is a change in the component state, React re-renders the component.

This change also impacts how the component behaves and renders on the screen. There are factors that can make the state change – for instance, a response to a user's action or system-generated events. Props and state are twin features of React. While props essentially pass information from a parent component to a child component, state alters components' internal data.

Let's take a look at a search use case for the implementation of state in components. Anytime a user types something into an HTML search input field, the user intends to see this typed information, which represents a new state, displayed somewhere else in the application.

The default state is the blank search input field. Hence, we need a way to change information in the input field and notify React to re-render its component. This means displaying the new state of the component.

Let's add a new component named `SearchSpeaker.js` to `src/SearchSpeaker/`:

```
import React,{useState} from "react";

const SearchSpeaker = () =>{
    const [searchText, setSearchText] = useState('');

    return (
        <div>
            <label htmlFor="search">Search speaker:
                </label>
            <input id="search" type="text" onChange={e =>
                setSearchText(e.target.value)} />

            <p>
                Searching for <strong>{searchText}
                    </strong>
            </p>
        </div>
    );
}

export default SearchSpeaker;
```

In the preceding code snippet, the `SearchSpeaker` component has two variables:

```
const [searchText, setSearchText] = useState('');
```

The `searchText` and `setSearchText` variables manage how the `searchSpeaker` component updates its state upon changes in state. This line of code is an example of the array destructuring we discussed earlier, where we assign the values returned by the `useState` hook to two variables in a single line. In *Chapter 3, Managing State with React Hooks*, we will discuss `useState` extensively.

The `searchText` variable is used to set the current state and tells React to re-render its `searchSpeaker` component whenever the event handler notifies a change in state with the new value or state set by `setSearchText`.

In React, `useState` is a utility function that takes an initial state as an argument. In this case, the initial state is empty: `useState('')`. The empty string initial state notifies React that the state will change over time. Therefore, `useState` comprises a two-entry array. The first entry, `searchText`, represents the current state; the second entry is a function to change the state using `setSearchText`.

These are the mechanisms we use to display the current state or change the state inside a React component.

In `App.js`, add the following code:

```
...
import SearchSpeaker from './SearchSpeaker';

function App()
{
    return (
        <div style={{backgroundColor: 'blue', margin:20,
            color:'white'}}>
            <h1>...</h1>
            .....
            <SearchSpeaker    />
        </div>
    );
}

export default App;
```

The following screenshot shows how the state changes in a React component:

Figure 2.11 – Screenshot showing the onChange state of the form field

With the understanding of state as an object used by React components to hold and manage data that can change over time, developers can create interactive and dynamic UIs that respond to user actions, and also provide a better user experience.

Summary

In this chapter, we discussed some of the core concepts in React applications. We started with the basic anatomy of a React project structure as generated by the Create React App tool. Some of the purposes of the files and folders were explained. We discussed some of the ES6 features, such as using arrow functions, destructuring, and default and named exports.

We also defined components as a core building block of any React application. Two types of components were discussed: `class` and `function` components. In addition, we discussed props and how to pass information in props. The unidirectional information flow in React was clarified. Finally, we discussed state as a way React manages the internal data.

In the next chapter, we'll dive deeper into React application development by discussing some of the hooks available in React. This will expose us to some of the advanced topics in React.

3

Managing State with React Hooks

Chapter 2, Getting Started with React, was a great way to kick off React frontend development. By now, you should be familiar with project directory structures and a few other concepts in React. In this chapter, we will take your understanding of React's core concepts further.

Why does this matter? Simple. You can't be the shining light you intend to be with React development without getting a grounding in the React core features and how we use them. This chapter focuses on managing state with React **Hooks**.

State in React is the medium through which we add interactivity to the user interface. Before React v16.8, developing class components was the only way you could add state and state transitions to your components.

Functional components were stateless; they were only able to display **JavaScript XML (JSX)** elements, that is, presentational components only. But with the Hooks API, you can add state and state transitions to your functional components.

In this chapter, you will come to understand various React hooks and how we use them to add statefulness to functional components. The building blocks of any React application are components, and making them stateful is what enhances the user experience of web applications.

By the end of this chapter, you will be able to build stateful function components with hooks such as `useState`, `useEffect`, `useContext`, `useMemo`, and `useReducer`, and even be able to develop your own custom Hooks.

In this chapter, we will be covering the following topics:

- What is a Hook in React?
- Why use Hooks in React?
- Using `useState` to develop stateful components

- Using `useEffect` to create useful side effects
- Using `useContext` to manage React applications' global state
- Using `useRef` to directly access DOM elements and persist state values
- Using `useReducer` for state management
- Using `useMemo` to improve performance
- Using `useCallback` to avoid re-rendering functions
- Using custom Hooks for code reusability

Technical requirements

The complete code for this chapter is available on GitHub at: `https://github.com/PacktPublishing/Full-Stack-Flask-and-React/tree/main/Chapter03`.

Due to page count constraints, the code blocks have been snipped. Please refer to GitHub for the full source code.

What is a Hook in React?

A **hook** is a special function provided by React that lets you use React core features—state and component lifecycle methods—within a function component. While state is an in-built object in React that adds interactivity and dynamic mechanism to components, the lifecycle tracks the phases components go through, from their initialization to their eventual demise (when a user navigates away or exits from an application UI).

There are three major cyclic phases React components go through, as explained in *Chapter 2, Getting Started with React*: mounting, updating, and unmounting. Each of these phases has what we call lifecycle methods that can be used during the rendering of React components.

We observed the presence of certain methods, such as `componentWillMount()`, `componentDidMount()`, `componentWillUpdate()`, and `componentDidUpdate()`, during the class component's lifecycle. React hooks are used to make function components stateful without using class components' lifecycle methods.

If you were working with stateful components before React version 16.8, you had no choice but to use class components as the only way to incorporate statefulness into your components.

Let's look at a component that changes a first name to a full name when a button is clicked:

```
import React from 'react';
class App extends React.Component{
  constructor(props) {
    super(props);
```

```
      this.state = {
        name: "Andrew",
      }

      this.updateNameState = this.updateNameState.bind(this);

    }
    updateNameState(){
      this.setState({
        name: "Andrew Peter"}
  );
    }
    render() {
      return(
        <div>
          <p>{this.state.name}</p>
          <button onClick={this.updateNameState}>Display Full
            Name</button>
        </div>

      );
    }
}
export default App;
```

Let us understand the preceding code in detail:

- `import React from 'react'`: This line brings React library core features into the scope.

- `class App extends React.Component`: This declares our `class App`, which extends the React component base class.

- The following snippet defines a constructor that accepts `props` as an argument:

  ```
  constructor(prop) {
    super(props);
    this.state = {
      name: "Andrew",
  }
  ```

This is a normal JavaScript class construct. Any class that extends the base class must have a `super()` method defined. The `this.state={name:"Andrew",}` part sets the initial state to `Andrew`. This is the state we want to update later in the code.

- The following snippet ensures that the function's `this` context will refer to the correct instance of the component when called:

```
this.updateNameState = this.updateNameState.bind(this);
```

  The `updateNameState` function is bound to the component instance using `.bind(this)`.

- The following snippet demonstrates the state updater method:

```
updateNameState(){
  this.setState({
    name:"Andrew Peter"
  });
}
```

  It is invoked in our button to set state from `name:"Andrew"` to `name: "Andrew Peter"`.

- `render()`: This is a compulsory method for every class component in React.

- `<p>{this.state.name}</p>`: This sets our initial state, which is `Andrew`, and returns it for our viewing as JSX.

- According to the following snippet, when the button is clicked, the `updateNameState()` class method is invoked and set to an updated state, which is `Andrew Peter`:

```
<button onClick={this.updateNameState}>
  ChangeToFullName</button>
```

In this chapter, we will refactor the preceding code snippet to a function component using a Hook. But before we delve into this, let's look at two rules guiding how we write Hooks in React function components:

- **Rule 1**: *Hooks must only be invoked at the top level.*

  You can't call Hooks from inside conditions, loops, or nested functions. Rather, you are to always invoke Hooks at the top level of your React function.

- **Rule 2**: *Hooks must only be invoked from a React component function.*

  You can't invoke Hooks from regular JavaScript functions, and neither can you invoke Hooks from the class component in React. You can only invoke Hooks from functional components. You can also invoke Hooks from custom Hooks.

Why use Hooks in React?

In the history of React, Hooks represented a significant shift in how we approach stateful components and manage side effects. Prior to Hooks, writing or refactoring class components was the primary method to enable components to exhibit interactivity and handle other side effects. Components serve as the building blocks of React applications' UIs, and creating interactive interfaces necessitated the use of class components.

However, for beginners, the class syntax and structure can be challenging to understand. Sophie Alpert, former manager of the React team at Facebook, in her keynote (*React Today and Tomorrow*) at the *2018 React Conference*, said:

"I claim classes are hard for humans…but it's not just humans, I claim the classes are also hard for machines"

– Sophie Alpert (https://bit.ly/37MQjBD)

The use of `this` and `bind` in class components adds to the list of confusion. While JavaScript offers both the world of **Object-Oriented Programming (OOP)** and that of functional programming, with React class components, you can't code without understanding the OOP paradigm.

This critically underscores the challenges that newcomers to the React world face. This was at least the case until React Hooks came into the picture. With Hooks, you simply write regular JavaScript functions that are easier to code, and you just have to hook into React Hooks for statefulness.

Another reason you might opt for React Hooks is the reusability of stateful logic across multiple components. Hooks allow you to separate stateful logic from component rendering logic, making it easier to reuse the logic in different components.

This separation ensures better modularity and reusability, as you can share your custom Hooks containing stateful logic across different React applications and with the broader React community. On the other hand, with class-based components, stateful logic and UI are often intertwined, which can make it harder to extract and reuse the logic efficiently.

In sum, React Hooks have triggered a new way of thinking about React component design. The possibility of gradual adoption in an existing code base (if you are still running on legacy React source code), makes it easy for diehard class component React developers to continue to write their stateful class components along with systematic migration of their code base to a functional-oriented approach.

The future of React is pointed toward functional component architecture. I can't imagine anyone reasonably pursuing class components anymore at this stage. By learning about and writing function components, developers can harness the advantages of React more effectively.

In the next section, we will start the process of developing a stateful React component using Hooks. We will start with the `useState` Hook, the most popular React Hook, which brings state to function components.

Using useState to develop stateful components

The `useState` Hook allows you to manage state in React applications. Function components rely on the `useState` Hook to add state variables to them. State is an object in React that can hold data information for use in React components. When you make a change to existing data, that change is stored as a state.

This is how it works: you pass an initial state property to useState(), which then returns a variable with the current state value and a function to update this value. The following is the syntax of the useState Hook:

```
const [state, stateUpdater] = useState(initialState);
```

Let's see a simplistic use case of how useState works:

```
import React, {useState} from 'react';
const App = () => {
  const [count, setCount] = useState(0);
  const handleIncrementByTen = () => {
    setCount(count + 10);
  };
  const handleDecrementByTen = () => {
    setCount(count - 10);
  };
  const resetCountHandler = () => {
    setCount(0)
  };
```

The preceding code snippet shows how you can develop a component that has increment, decrement, and reset states. When the IncrementByTen button is clicked, the counter increases the number by 10 and when the DecrementByTen button is clicked, the decrement state is activated and the number decreases by 10.

The reset to the initial state does what it's meant to do – it resets the value to its initial value. The following completes the code snippet:

```
return (
  <div>
    Initial Count: {count}
    <hr />
    <div>
      <button type="button"
        onClick={handleIncrementByTen}>
        Increment by 10
      </button>
      <button type="button"
        onClick={handleDecrementByTen}>
        Decrement by 10
      </button>
      <button type="button" onClick={resetCountHandler}>
        Reset to Initial State
      </button>
```

```
        </div>
      </div>
    );
  };
export default App;
```

Let's understand the preceding code in more detail:

- Importing useState: To use the useState Hook function, we first need to import it into our component from the Import React, { useState } from 'react' React object.

- Initializing useState: We initialize our state by calling useState in our component as follows:

  ```
  const [count, setCount] = useState(0);//using destructuring
  array to write a concise code.
  ```

  useState<number> accepts an initial state of zero (useState(0)) and returns two values: count and setCount:

  - count: The current state

  - setCount: State updater function (this function is responsible for the new state of the initial state)

- useState(0): useState <number> with the initial value of 0

  In useState, you can only declare a state property at a time. However, the data can be of any type: primitives, arrays, and even objects.

- **Reading state**: The following part of the code lets us use the state properties in the rendered component:

  ```
  <div>
        Initial Count: {count}
        <hr />
        <div>
          <button type="button"
            onClick={handleIncrementByTen}>
            Increment by 10
          </button>
          <button type="button"
            onClick={handleDecrementByTen}>
            Decrement by 10
          </button>
          <button type="button"
            onClick={resetCountHandler}>
            Reset to Initial State
          </button>
  ```

```
            </div>
          </div>
```

The onClick event functions are added to help emit event operations for our buttons. When the buttons are clicked, different event functions are invoked based on the expected actions.

- **Update state**: We use the updater function to change the value of state. The handleIncrementByTen(), handleDecrementByTen(), and resetCountHandler() functions are used to change the state values, as shown in the following snippet:

```
const handleIncrementByTen = () => {
  setCount(count + 10);
};

const handleDecrementByTen = () => {
  setCount(count - 10);
};
const resetCountHandler = () => {

  setCount(0)
};
```

useState<number> can contain primitive and object data that can be accessed across React components. At this point, it is recommended you fire up your VS code or your preferred IDE and experiment with useState in developing a stateful component.

Passing state as props

State is not limited to being used solely within the component where it is defined. You can pass state as props to child components, allowing them to display or use the parent state data.

Let's consider the following example:

```
import React, { useState } from 'react';
const ParentComponent = () => {
  const [count, setCount] = useState(0);
  const handleIncrementByTen = () => {
    setCount(count + 10);
  };
  return (
    <div>
      <p>Parent Count: {count}</p>
      <ChildComponent count={count} />
```

```
      <button onClick={handleIncrementByTen}>Increment
      </button>
    </div>
  );
};

const ChildComponent = ({ count }) => {
  return <p>Child Count: {count}</p>;
};
```

The preceding code shows a React function component with state using the useState Hook. It consists of two components, ParentComponent and ChildComponent, and demonstrates how to pass state data from the parent component to the child component.

When you use ParentComponent in your application, it will render with an initial count : number of 0. **Parent Count** will display the current value of parent count : number, and **Child Count** (rendered by ChildComponent) will also display the same value as it receives it via the count prop. When you click the **Increment** button, the count state will increase by 10, and both counts will reflect the same updated value.

Conditional rendering with state

Conditional rendering with state in React allows you to show or hide specific parts of the user interface based on the values of state variables. By using conditionals, you can control what content or components are displayed depending on the current state of your application.

This technique can be useful for creating dynamic and interactive user interfaces. Imagine having a **Login** button that turns into a **Logout** button once the user is logged in. That's a classic example of conditional rendering! When you click the button, React will automatically update the UI to reflect the new state, making it super responsive. Oh, and that's not all!

You can even use this magic to toggle the visibility of different elements, such as showing or hiding a cool modal or drop-down menu based on user actions. For instance, let's say you have the isLoggedIn state variable, and you want to display different content based on whether the user is logged in or not.

The following code demonstrates how you can implement this with the useState Hook:

```
import React, { useState } from 'react';
const Dashboard = () => {
  const [isLoggedIn, setIsLoggedIn] = useState(false);

  const handleLogin = () => {
    setIsLoggedIn(true);
  };
```

```
const handleLogout = () => {
  setIsLoggedIn(false);
};

return (
  <div>
    {isLoggedIn ? (
      <button onClick={handleLogout}>Logout</button>
    ) : (
      <button onClick={handleLogin}>Login</button>
    )}

    {isLoggedIn && <p>Hey friend, welcome!</p>}
    {!isLoggedIn && <p>Please log in to continue.</p>}
  </div>
);
};
```

The preceding code demonstrates a React component called `Dashboard`. It's all about handling user authentication and showing personalized messages to the user.

Inside the `Dashboard` component, we have the `isLoggedIn` state variable, which is managed using the `useState` Hook. This variable keeps track of whether the user is currently logged in or not. When the component first renders, the initial state of `isLoggedIn` is set to false, indicating that the user is not logged in.

Now, let's dive into the magic of conditional rendering! When you look at the JSX inside the `return` statement, you'll see some interesting stuff happening. We use the `{}` curly braces to wrap our conditionals.

If `isLoggedIn` is `true`, we display a **Logout** button, and if it's false, we show a **Login** button. Depending on the state, the appropriate button will be rendered, and each button has an `onClick` event that triggers the respective `handleLogin` or `handleLogout` function. The fun doesn't end there!

We also use more conditional rendering with `isLoggedIn` to display a personalized message for the user. When `isLoggedIn` is `true`, we show a warm greeting such as **Hey friend, welcome!**, and when it's false, we kindly ask the user to log in to continue. So polite, right? When the user clicks the **Login** button, the `handleLogin` function gets called, and guess what?

It sets `isLoggedIn` to `true`, indicating that the user is now logged in! Likewise, when the user clicks the **Logout** button, the `handleLogout` function is triggered, and it sets `isLoggedIn` back to `false`, meaning the user is now logged out.

In the next section, we will examine another Hook in React, `useEffect`. It is widely used to manipulate the DOM and fetch data from external sources.

Using useEffect to create side effects

The useEffect Hook allows you to fetch data from external sources, update the DOM tree, and set up a data subscription. These operations are called side effects. In the class component, you have what we call lifecycle methods that can execute operations based on the phase of the component-rendering process. useEffect accepts two arguments: a function and an optional dependency.

It is important to note that useEffect does the work of the old componentDidMount, componentDidUpdate, and componentWillUnmount in one place. Using the useEffect Hook shortens the amount of code you have to write in a function component to achieve the same side effects.

The following is the syntax for the useEffects Hook:

```
- useEffect(<function>, <dependency>)
useEffect(() => {
  // This callback function implementation is either to
     update DOM, fetch data from external sources, or to
     manage subscription that happens here.
}, [dependency]);
```

Let's dive into an example of using the useEffect Hook:

```
import React, { useEffect, useState } from 'react';
const App = () => {
const [data, setData] = useState([]);
    const API_URL = "https://dummyjson.com/users";
      useEffect(() => {
        fetchSpeakers();
    }, []);
     return (
        <ul>
       {data.map(item => (
         <li key={item.id}>
            {item.firstName} {item.lastName}
         </li>
       ))}
      </ul>);
};
export default App;
```

In the preceding code, we are fetching data from an external API source using useEffect. For this example, we have used fake API data from https://dummyjson.com/users. By the time we get to the backend development section of this book (*Chapter 9, API Development and Documentation*), we will be developing custom API endpoints. Next, we will use the useEffect() Hook to call the fetchSpeakers function.

Refer to GitHub for the full code and have a look at the following points:

- `import React, { useEffect, useState } from 'react';`: This line allows us to use `useEffect` and `useState` APIs from the React library.

- `const [data, setData] = useState([]);`: In this line, we have declared a state object set with an empty array as the initial data.

- `useEffect(()=>{...}, [])`: This part of the code represents a signature set that fetches data from the specified external source. The second argument in the `useEffect` function, the dependency `[]`, is set to an empty array. The empty array ensures `useEffect()` renders just once, the first time, on mounting.

  To have it render depending on state changes, you would have to pass the state through the dependency array. With this, you are able to prevent the constant unnecessary re-rendering of the component unless the dependency state changes.

- `fetchSpeakers():Promise<Speaker[]>` inside the `useEffect` Hook is a call to the `fetchSpeakers` function. This function is an asynchronous function that fetches data from a mocked remote API and sets the data in the component's state using the `setData` function. The empty dependency array `[]` passed as the second argument to `useEffect` indicates that the effect should only run once when the component is mounted and never again after that.

  Since there are no dependencies listed, the effect won't be triggered by changes in any props or state variables. This is why it behaves like the `componentDidMount` lifecycle method, as it runs only once when the component is first rendered.

- `const API_URL = "https://dummyjson.com/users";`: The `API_URL` variable is set to hold the endpoint information about the external source.

- The `try... catch` code block is set to execute the code and console error if there is an error fetching data from the endpoint:

```
const fetchSpeakers = async () => {
  try {
    const response = await            fetch(API_URL);
    const data = await response.json();
    setData(data.users);
  } catch (error) {
    console.log("error", error);
  }
};
```

This preceding code snippet used the web browser `fetch()` API to fetch the data from the `API_URL` endpoint. The `try... catch` code block is set to execute the code and console error if there is an error fetching data from the endpoint.

- `map()` is set on the data to loop through the object array data and display the newly created array from the function invocation on every array element `item: Speaker`:

```
{data.map(item => (
        <li key={item.id}>
          {item.firstName} {item.lastName}
        </li>
    ))}
```

Let's update the `useEffect` Hook function of the preceding code and add a state to its dependency and a `Cleanup` function. Adding a `Cleanup` function inside the `useEffect` hook serves a crucial purpose in React applications.

The `cleanup` function is executed when the component unmounts or when the dependencies listed in the `useEffect` hook change. Its main use is to perform cleanup tasks, freeing up resources, and preventing potential memory leaks or unexpected behavior in the application.

Now, update the preceding `useEffect()` as follows:

```
useEffect(() => {
    const fetchData = async () => {
      const fetchedData = await fetchSpeakers();
      if (isMounted) {
        setData(fetchedData);
      }
    };
    fetchData();
    // Cleanup function
    return () => {
      isMounted = false;
    };
  }, [data]) ;// Adding data state as a dependency
```

The preceding code uses the `useEffect` Hook to fetch data from an API (`fetchSpeakers`) and update the data state with the fetched result. It employs an `isMounted` flag to prevent setting the state after the component has unmounted, effectively avoiding potential issues. The data fetching effect runs whenever the `data` state changes, and the `Cleanup` function sets the `isMounted` flag to false when the component unmounts.

In sum, we have seen how `useEffect` could be used for side effect operation in a function component by fetching data from an external source. Next, we will look at how we can use the `useContext` Hook to better manage the global state in React applications.

Using useContext to manage global state in React applications

The useContext Hook is used to share application state data across the component tree without having to pass props down explicitly at every component level. To put it simply, useContext is a way to manage React applications' global state. Remember, we used the useState Hook to manage local state in the *Using useState to develop stateful components* section.

However, as React project requirements expand in scope, it will be ineffective to use the useState Hook alone in passing state data in deeply nested components. The following is the syntax for the useContext Hook:

```
const Context = useContext(initialValue);
```

Briefly, we will discuss *props drilling* to understand the challenges it poses. Afterward, we'll delve into the implementation of the context API, which addresses these issues.

Understanding props drilling

Let's examine how you might pass data as props down a component hierarchy without the use of useContext. The following code snippet shows how we pass data to inner deeply nested components without the use of useContext:

```
  import React, {useState } from 'react';
const App = () => {
  const [speakerName]= useState("Fred Morris");
  return (
    <div>
      <h2>This is Parent Component</h2>
      <ImmediateChildComponent speakerName={speakerName} />
    </div>
    );
    }
    function ImmediateChildComponent({speakerName}) {
      return (
        <div>
          <h2>This is an immediate Child
            Component</h2><hr/>
          <GrandChildComponent speakerName={speakerName} />
        </div>
      );
    }
```

```
    }
export default App;
```

The preceding code displays the name of a speaker in a function comprising nested components. The full source code is on GitHub.

Let us understand the code in more detail:

- `const [speakerName] = useState`: This line is used to set the default state for `speakerName`.

- `<App />` is a parent component that passes state using `{speakerName}` as props for the state needed in `<GrandChildComponent />`:

```
      return (
        <div>
          <h2>This is Parent Component</h2>
          <ImmediateChildComponent
            speakerName={speakerName}
        </div>
        );
```

The parent has to pass through the `<ImmediateChildComponent />` component to reach `<GrandChildComponent />` nested lower in the hierarchy. This becomes even more cumbersome when you have five or more intermediary components before we get to the actual component that needs the state information.

This is the problem `useContext` tries to solve. The following code shows the intermediate component and the final `GrandChildComponent`: `React.FC<Props>` where the states are required:

```
function ImmediateChildComponent({speakerName}) {
    return (
      <div>
        <h2>This is an immediate Child
          Component</h2><hr/>
        <GrandChildComponent
          speakerName={speakerName}   />

      </div>
    );
  }

function GrandChildComponent({speakerName}) {
    return (
      <div>
        <h3>This is a Grand Child Component</h3>
```

```
            <h4>Speakers Name: {speakerName}</h4>

        </div>
    );
}
```

Let's now look at how `useContext` can be used to solve the preceding problem by maintaining a global state where different components can communicate without causing prop drilling issues in React.

Using useContext to solve the props drilling problem

With `useContext`, you will understand how you can pass state data across components without manually doing so with props. The following code shows how `useContext` is used:

```
import React, {useState, useContext,createContext } from
  'react';
const context = createContext(null);
const App = () => {
const [speakerName]= useState("Fred Morris");
  return (
    <context.Provider value={{ speakerName}}>
            <h1>This is Parent Component</h1>
            <ImmediateChildComponent  />
      </context.Provider>
        );}
function ImmediateChildComponent() {
    return (
      <div>
        <h2>This is an immediate Child Component</h2>
        <hr/>
        <GrandChildComponent  />
      </div>);
}
}
  export default App;
```

Let us understand the preceding code in detail. Refer to GitHub for the full source code:

- `import React, {useState, useContext,createContext } from 'react';`: This line allows us to make use of `useState`, `useContext`, and `createContext` from the React library.

- `const context = createContext(null);`: This line creates `Context<ContextType>` and allows us to use `Provider: React.FC<ProviderProps|null>` with `null` as the initial value. Note that the `null` default value could also be any value provided to us by the `createContext` function.

- The context provider envelopes the child component and makes the state values available, as follows:

```
return (
  <context.Provider value={{ speakerName }}>
    <h1>This is Parent Component</h1>
    <ImmediateChildComponent />
  </context.Provider>
);
```

- `const {speakerName} = useContext(context);`: In this line, we use the `useContext` Hook to have access to `context` in `<GrandChildComponent />`:

```
function GrandChildComponent():React.FC<Props> {
  const {speakerName} = useContext(context);
    return (
      <div>
        <h3>This is a Grand Child Component</h3>
        <h4>Speaker's Name: {speakerName}</h4>

      </div>
    );
```

In sum, the `useContext` Hook enables us to use `context` in function components no matter how nested the component hierarchy may be. This is always required in complex React applications where the state data may be needed across global application states. With `useContext`, we are able to share the information state that was passed as props without the direct interference of the intermediate components.

Next, we will delve into the `useRef` Hook and explore how it can be effectively utilized in a React component.

Using useRef to directly access DOM elements and persist state values

The `useRef` Hook allows you to access DOM elements directly in React and is used to persist state values across re-renders. React, as a powerful library for UI, has a lot of novel concepts (virtual DOM design patterns, event handling, attribute manipulation) we can use to access and manipulate DOM elements without the use of traditional DOM methods.

This declarative approach to DOM is one of the reasons React is so popular. However, with useRef, we can directly access DOM elements and freely manipulate them without consequence. The team at React felt that with useRef, developers' present and future cravings for direct DOM access might be met despite the React abstraction on top of the DOM.

There are two core uses of useRef:

- Accessing DOM elements directly
- Persisting state values that do not trigger the re-rendering of React components when updated

If you are interested in how many times a component re-renders upon update, we can use either useState or useRef. But it will be a bad idea to use useState. Using this might leave users stuck in an infinite loop of re-rendering since useState re-renders on every update of its values.

However, the useRef Hook shines in this scenario as it can store state values across components re-rendering without triggering the re-render mechanism.

Let's dive into a use case for useRef in the form of autofocusing an input field on a rendered component:

```
import React, {useRef} from 'react';
const App = () => {

    const inputRef = useRef(null);
    const clickButton = () => {
      inputRef.current.focus();
    };
    return (
      <>
        <input ref={inputRef} type="text" />
        <button onClick={clickButton}>click to Focus on
          input</button>
      </>
    );
  }
export default App
```

Let us understand the preceding code in detail:

- const inputRef = useRef(null);: This line creates a reference for the useRef function
- <input ref={inputRef} type="text" />: In this line, ref is added to the input element to make use of the useRef() Hook
- The onClick event is added to the button, which makes use of inputRef <button onClick={clickButton}>click to focus on input</button>

useState and useRef are very similar in the sense that they hold state values. However, useState re-renders each time its values change, while useRef does not trigger a re-render.

Let's move on to the next Hook, useReducer. It is another hook to manage complex states in React applications.

Using useReducer for state management

The useReducer hook is a state management hook in a React application. It is quite a bit more robust than the useState hook we discussed earlier in this chapter as it separates the state management logic in the function component from the component-rendering logic.

The useState hook encapsulates the state management function with component rendering logic, which may become complex to handle in a large React project with the need for complex state management. The following is the syntax for useReducer:

```
`const [state, dispatch] = useReducer(reducer, initialState)
```

The useReducer hook accepts two arguments – the reducer, which is a function, and the initial application state. The Hook then returns two array values – the current state and the Dispatch function.

Basically, we need to understand these core concepts in useReducer:

- State: This refers to mutable data that can be changed over time. State doesn't have to be an object; it could also be an array or number.
- Dispatch: This is a function that allows us to modify the state. Dispatch is used to trigger the action that changes the state.
- Reducer: This is a function that handles the business logic of how the state could be modified.
- IntialState: This refers to the initial state of the applications.
- Action: This is an object with a set of properties. Type is a required property.
- Payload: This refers to the data of interest in a chunk of network data.

With these core concepts explained, we also need to understand one more thing: the main purpose of useReducer is to manage complex multiple states in such a way that the logic for state management is separated from the component view functionality. We will elaborate more on this with a practical example.

Let's dive in to see the use case for useReducer:

The following snippet will show how you can use useReducer to manage different state properties. We will be working with an event schedule component. In the following code snippet, we are fetching the data from fake JSON API data.

In the `src` directory, create `src/db.json` and paste in this data object:

```json
{    "schedules":
    [
        {
            "id":1,
            "time":"10.00 AM",
            "speaker": "Juliet Abert",
            "subjectTitle":"Intro to React Hooks",
            "venue":"Auditorium C"
        },
        {
            "id":2,
            "time":"12.00 AM",
            "speaker": "Andrew Wilson",
            "subjectTitle":"React Performance Optimization"
            ,"venue":"Auditorium A"
        },
        {
            "id":3,
            "time":"2.00 PM",
            "speaker": "Lewis Hooper",
            "subjectTitle":"Intro to JavaScript",
            "venue":"Auditorium B"
        }

    ]
}
```

To install the JSON server for mocking backend services, in the terminal, enter the following command:

```
npm i -g json-server
```

Start the server on port 8000 with the following command:

```
json-server --watch db.json --port=8000
```

Once the JSON server is started, the following will appear in your terminal:

```
Loading db.json
  Done
  Resources
  http://localhost:8000/schedules
  Home
  http://localhost:8000
```

Add the following snippet to App.js:

```
import { useReducer, useEffect } from 'react';
import axios from "axios";
const initialState = {
  isLoading: false,
  error: null,
  data: null,
};
const reducer = (state, action) => {
  switch (action.type) {
    case "getEventSchedule":
      return {
        ...state,
        isLoading: true,
        error: null,
      };
              </ul>
    </div>
  );
};

export default App;
```

The full source code can be found on GitHub. Let's examine the code snippet:

- The initial state properties of the component are first specified:

```
const initialState = {
isLoading: false,
error: null,
data: null,
};
```

- Then, we define the Reducer function as const reducer = (state, action) => {}.

  The Reducer function takes two arguments: state and action. Then the action through the type property defines the logic of the state. In this case, the switch runs through a series of conditional action-based operations and returns a specific action type.

  The action-type properties specified in the reducer function, for instance, getEventSchedule, getEventScheduleSuccess, and getEventScheduleFailure, allow us to modify the state of the component based on the state of the action type.

- getEventSchedule<EventSchedule[]> accepts all the properties of initalState, and the isLoading property is set to true because we are fetching this state data:

```
case "getEventSchedule":
     return {
       ...state,{/*accepts other initial State
         properties*/}
       isLoading: true, {/*change the initial state
         of isLoading*/}
     };
```

- getEventScheduleSuccess will be invoked when the data property is modified through the returned data in action.payload: EventSchedule[], and the isLoading property is set back to false:

```
  case "getEventScheduleSuccess":
     return {
       ...state,
       isLoading: false,
       data: action.payload,{/*we have useful
         returned data at this state*/}
     };
```

And if there is no returned data, getEventScheduleFailure:Action is invoked and an error is displayed:

```
.catch(() => {
  dispatch({ type: "getEventScheduleFailure" });
});
```

- The App() component handles the view part of the component state where useReducer() is defined and executed:

```
const [state, dispatch] = useReducer(reducer, initialState);
```

The useReducer() accepts two arguments—reducer and initialState—and returns two array variables: state and dispatch. The state variable holds the state object and dispatch is a function that allows the reducer to update the state based on the invoked type of action in the reducer function.

- useEffect() is invoked to fetch the schedule data from the endpoint specified:

```
useEffect(() => {
  dispatch({ type:"getEventSchedule" });
  axios.get("http://localhost:8000/schedules/")
    .then((response) => {
      console.log("response", response);
      dispatch({ type: "getEventScheduleSuccess",
```

```
    payload: response.data });
})
```

Inside the useEffect() body, dispatch() is triggered based on the type of action. The object type is specified: dispatch({ type:"getEventSchedule" });.

- axios() is invoked to fetch the endpoint data with axios.get("http://localhost:8000/schedules/").

When the type of action is getEventScheduleSuccess, we expect returned data, thus the payload property – dispatch({ type: "getEventScheduleSuccess", payload: response.data }).

The following snippet handles the error that may occur from this promise-based request:

```
.catch(() => {
  dispatch({ type: "getEventScheduleFailure" });
});
```

In the App() component return construct, we render schedules to the screen with the following snippet:

```
<h2>Event Schedules</h2>
{state.isLoading && <div>Loading...</div>}
{state.error && <div>{state.error}</div>}
{state.data && state.data.length === 0
   &&   <div>No schedules available.</div>}
<ul>
   {state.data && state.data.map(({ id, time,
      speaker, subjectTitle, venue }) => (
      <li key={id}>
        Time: {time} <br />
        Speaker: {speaker}<br />
        Subject: {subjectTitle}<br />
        Venue: {venue}
      </li>
   ))}
</ul>
```

We check whether initialState :State is in the loading state and display <div>Loading...</div>. If the error state is true, we display the error. If there is no data to fetch, we display the appropriate message. We also check the data state and ensure we have data to display. Now, start the server if it's not running, with npm start:

The following screenshot shows an example implementation of useReducer.

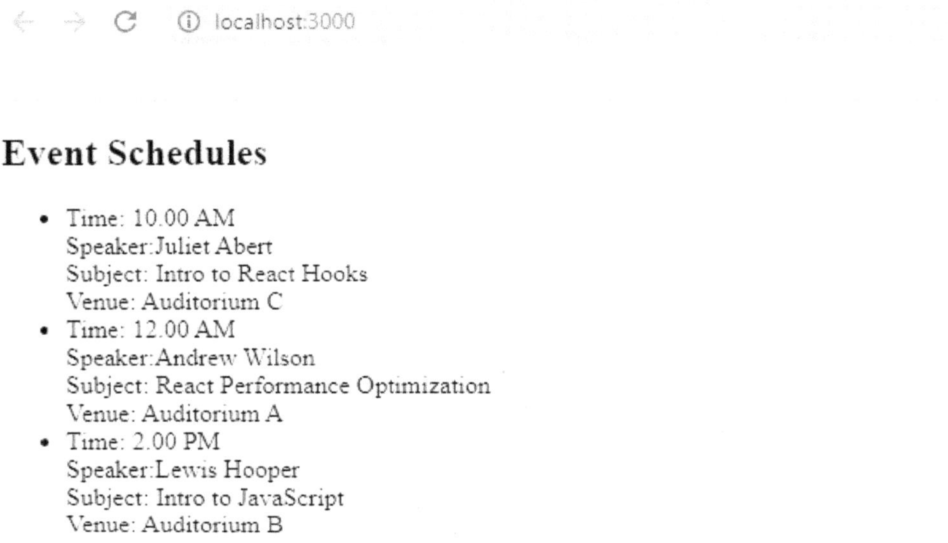

Event Schedules

- Time: 10.00 AM
  Speaker:Juliet Abert
  Subject: Intro to React Hooks
  Venue: Auditorium C
- Time: 12.00 AM
  Speaker:Andrew Wilson
  Subject: React Performance Optimization
  Venue: Auditorium A
- Time: 2.00 PM
  Speaker:Lewis Hooper
  Subject: Intro to JavaScript
  Venue: Auditorium B

Figure 3.1 – Screenshot showing the effect of the useReducer Hook

We have seen how we can use the useReducer Hook to manage advanced multiple states in React. In the next section, we will examine what useMemo is and how we can use it to improve performance in React applications.

Using useMemo to improve performance

The useMemo Hook is part of the core APIs in React geared toward improving the performance of React applications. It uses a technique known in software development as **memoization**.

This is an optimization technique used to enhance the performance of software by keeping in memory the results of resource-intensive computation function calls and sending back the cached output when the same inputs are used subsequently. So, why is useMemo important in React application development? useMemo solves two performance problems for a React developer.

Prevents unnecessary component re-rendering

It memoizes the return value of a function for computations that consume a lot of resources by sending back a cached function result upon subsequent requests without a state update.

Let's dive into a use case for useMemo to better understand how it can be used in React. This snippet shows how a component re-renders on every character search. With a large application that has over 20,000 users, this could result in performance issues.

First, we'll see what the code looks like without useMemo:

```
import React, { useState} from 'react';
const speakers = [
  {id: 10, name: "John Lewis"},
  { id: 11, name: "Mable Newton"},
];
const App = () => {
  const [text, setText] = useState("");
  const [searchTerm, setSearchTerm] = useState("");
  const onChangeText = (e) => {
    setText(e.target.value);
  };
  console.log("Text", text);
  const handleClick = (e) => {
    setSearchTerm(e.target.value);
  };
  console.log("Search Term", text);
  ));
  });
  return (
      <div>
        ---
    </div>
  );
};

export default App;
```

The following screenshot shows the `list` component re-rendering on every character search:

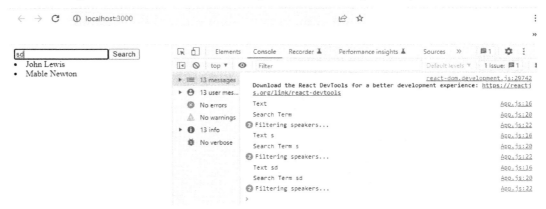

Figure 3.2 – A console showing components re-rendering

Let's go through the implementation of the `useMemo` Hook and gain insights into how developers can significantly enhance performance and optimize resource usage in React applications, ensuring that expensive computations are only executed when necessary:

- **Hypothetical data source**: `speakers` is declared to hold the array of object data:

```
const speakers = [
  {id: 10, name: "John Lewis"},
  { id: 11, name: "Mable Newton"},
];
```

- **Variables to state**: `text` and `searchTerm` are declared with their setter methods as state variables:

```
const [text, setText] = useState("");
const [searchTerm, setSearchTerm] = useState("");
```

- onChange handler: This event handler updates the initial state to the current state:

```
const handleClick = (e) => {
    setSearchTerm(e.target.value);
  };
```

- **Function to filter speakers**: The `filteredSpeakers` function is used to filter an array of speakers based on `searchTerm` using a case-insensitive search. With this filtering, you are able to optimize the performance of filtering by memoizing the filtered results:

```
const filteredSpeakers = speakers.filter((speaker) => {
  console.log("Filtering speakers...");
```

```
        return speaker.name.toLowerCase()
          .includes(searchTerm.toLowerCase());
    }
```

- **Output to screen**: This shows the result of rendering the component with the effect of useMemo:

```
<div>
  <input type="text" onChange={onChangeText} />
  <button onClick={handleClick}>Search</button>
</div>
{filteredSpeakers.map((filteredSpeaker) => (
  <li key={filteredSpeaker.id}>
    {filteredSpeaker.name}</li>
))}
</div>
```

As you can see in the preceding snippet, the dependency property for the speaker component hasn't changed. There is no need for re-rendering, but the console shows us that there is re-rendering.

Now let's see what the code looks like with useMemo. Update the filteredSpeakers function in the preceding code with the following snippet:

```
const filteredSpeakers = useMemo( () =>
  speakers.filter((speaker) => {
  console.log("Filtering speakers...");
  return speaker.name.toLowerCase()
    .includes(searchTerm.toLowerCase());
}, [searchTerm]));
```

The preceding snippet shows the use of useMemo on the filteredSpeakers function. This function only executes once the searchTerm state changes. The filteredSpeakers function is not expected to run when the text state changes, because that is obviously not a dependency in the dependency array for the useMemo Hook.

Next, we will explore useCallback. The useCallback Hook is similar to the useMemo Hook in making React applications performant. Both useCallback and useMemo optimize React applications. Let's dive in to understand useCallback in avoiding the re-rendering of component functions.

Using useCallback to avoid re-rendering functions

In React function components, there is an additional optimization Hook called useCallback. It shares functionality with useMemo, with a slight difference in output behavior in terms of what is returned. In useCallback, a memoized function is returned, while useMemo returns memoized returned values of the function.

Like useMemo, useCallback is invoked when one of its dependencies is updated inside the function component. This ensures that the functional components are not necessarily re-rendered constantly. There are key highlights for useCallback:

- A memoized callback function is returned in useCallback. This improves the performance of the React application based on memoization techniques.

- The change in the dependencies of the useCallback Hook determines whether it will update or not.

Right now, let's dive into a simple use case of useCallback for deeper understanding.

The following snippet displays a list of speakers to simulate the high computation requirements to consider the usage of useCallback for performance optimization. Also, it is important to note that this illustration is by no means sufficient enough as a use case to demonstrate real-life performance bottleneck scenarios, but it comes in handy to explain this scenario.

Assuming we have a huge list of speakers components to handle displaying and searching for speakers, without the use of useCallback, we will discover for every character search in the input field, App, List, and ListItem will be re-rendered needlessly.

The complete snippet can be found at https://github.com/PacktPublishing/Full-Stack-Flask-and-React/blob/main/Chapter03/08 in the book's GitHub repository:

```
import React, {useState,useCallback} from 'react';
  const handleRemoveSpeaker = useCallback(
    (id) => setSpeakers(speakers.filter((user) =>
      user.id !== id)),
    [speakers]
  );
```

The preceding code snippet demonstrates useCallback. The code structure is essentially similar to useMemo, except that useCallback is wrapped around the function we intend to cache or memoize.

In the following figure, we see how the App, List, and ListItem components re-render with every character search in the search input box.

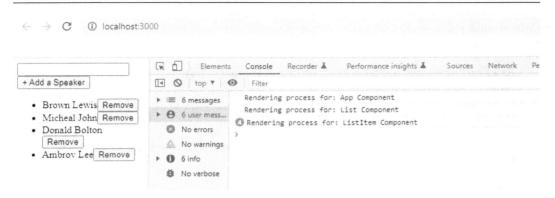

Figure 3.3 – Screenshot showing the usage of useCallback

`handleRemoveSpeaker: React.FC<ButtonProps>` is optimized with `useCallback` to prevent the re-rendering of the `List` and `ListItem` components due to a change in the state of the search input. The re-rendering of components is expected if users of the application click on the **Add a Speaker** or **Remove** buttons.

There is one major problem the `useCallback` Hook solves in React: preventing unnecessary re-rendering of components due to a referential equality check.

Next, we will dissect the use of custom Hooks to separate the business logic of the component from the rendering function. This allows function reusability and a better organization of code in a React application.

Using custom Hooks for code reusability

We have extensively discussed some of the in-built Hooks in React. Hooks have been part of the core React library since v16.8, which allows React components to exhibit statefulness without a class-based approach. Hooks such as `useState`, `useEffect`, `UseMemo`, `useRef`, `useContext`, and `useCallback` are specific functions to manage state, share stateful logic, and allow other interactions with React core APIs.

Now let's understand what a custom Hook is and the benefits we can get from using them.

Custom Hooks are normal JavaScript functions whose name starts with `use` and that usually invoke one or more in-built React Hooks. For instance, custom Hooks could be named anything as long as it starts with *use*, for instance, `useCustomHook`, `useFetchSpeakers`, or `useUpdateDatabase`. Conventionally, there must be `use` in front of your custom Hook name.

So why should you want to build your own custom Hooks? Let's examine some of the reasons experienced React developers build custom Hooks:

- As a React developer, you are going to write a ton of functions to solve problems in your React projects. And if the best React practices are not followed, some of those functions are going to be repeated so often in many components of your project. With a custom Hook, you can reuse stateful logic across several components in your project.

- Custom Hooks encourage the separation of concerns between state logic functions and view layers of components.

- Easy debugging.

Let's see an implementation example of custom Hooks:

Inside App.js, enter the following:

```
import React from 'react';
import useFetchSpeakers from "./useFetchSpeakers";
const App = () => {
  const API_URL = "https://dummyjson.com/users";
  const [data] = useFetchSpeakers(API_URL);
  return (
    <>
      <ul>
        {data.map((item) => (
          <li key={item.id}>
            {item.firstName} {item.lastName}
          </li>
        ))}
      </ul>
    </>
  );
};

export default App;
```

Now, let's break the preceding snippet down:

- `import useFetchSpeakers from "./useFetchSpeakers"` brings the custom Hook into scope in this application. Like any other Hook, we use *use* as part of the naming convention.

- The `useFetchSpeakers` Hook returns the data variable state, based on `API_URL :` `string` passed as the endpoint. This endpoint as an argument is passed to the custom `useFetchSpeakers` Hook.

- We then iterate over the data objects with map() to display the returned firstName:string and lastName:string.

Inside useFetchSpeakers.js, we define the custom Hook function with its locally managed state:

```
import { useEffect, useState } from 'react';

const useFetchSpeakers = (url) => {
  const [data, setData] = useState([]);
  useEffect(() => {
    const fetchSpeakers = async () => {
      try {
        const response = await fetch(url);
        const data = await response.json();

        setData(data.users);
      } catch (error) {
        console.log("error", error);
      }
    };

    fetchSpeakers();
  }, [url]);

  return [data];
};

export default useFetchSpeakers;
```

In the preceding snippet, the following steps were involved:

1. The useFetchSpeakers custom Hook signature is defined. It accepts url as an argument.

2. The useFetchSpeakers Hook uses useEffect() to asynchronously fetch data from an endpoint – the url argument passed into the custom Hook.

3. A promise is returned with *jsonified* result data, which is made available to the setData(data. users) state.

4. It has a dependency (url) that causes the component to re-render the component state upon any change in the data.

With this, you can see how the custom Hook allows the logic part of the component to be separated from the rendering part, and how code reusability is encouraged and implemented.

Summary

In this chapter, we have been able to understand Hooks as a new mind shift in how we add statefulness to components in React. Prior to Hooks, only class components could offer us stateful functionalities. With React 16.8, we are now able to develop stateful functional components in React applications that are more elegant and concise.

The learning curve is easy as we can leverage our understanding of regular JavaScript functions and develop function components to power user interfaces for our web applications. With Hooks in React, user and developer experiences have been greatly improved.

In the next chapter, we will focus extensively on how we can leverage React APIs to fetch data from external sources into our web applications. Most of the applications we use today rely heavily on external data. Undoubtedly, React shines well in this domain.

4
Fetching Data with React APIs

Over the past few years, there has been an increase in the demand for database-driven web applications. This increase is a consequence of the abundance of data available at this present time. With widespread internet adoption, businesses leverage web applications to interact with customers, employees, and other stakeholders.

More than ever, web developers are constantly faced with tasks such as the organization and consumption of data. Both internal and external data require us to have smart and business-oriented database-driven web applications.

As a full stack software engineer, some of your frontend tasks will be to consume data, either from an internally developed API or a third-party API. Before we delve into approaches or tools you can use to fetch data in React projects, let's briefly discuss what APIs are all about and why they are redefining ways of building user interfaces and web applications.

An API simply allows communication among systems using a set of rules in a standard accepted format. In web development, HTTP protocols define the set of rules for web-based systems communication. HTTP is a data exchange protocol used to fetch resources across the internet.

There are two major formats for exchanging data: **XML** and **JSON**. JSON is winning the popularity contest between these two widely used formats for data exchange. JSON is specifically designed for data interchange, handles arrays seamlessly, and is used widely among developers.

In the React ecosystem, developers have access to a range of exposed interfaces designed to facilitate data fetching from various sources. These APIs are aimed at empowering React developers to create intuitive user interfaces and enhance the overall user experience when interacting with web applications.

In this chapter, we are going to learn about some of the methods and techniques used in React frontend development to fetch data from disparate sources. In this chapter, we will be covering the following topics:

- Fetching data using the Fetch API in React
- Fetching data using `async/await` syntax

- Fetching data using Axios
- Fetching data using React Query

Technical requirements

The complete code for this chapter is available on GitHub at: `https://github.com/PacktPublishing/Full-Stack-Flask-and-React/tree/main/Chapter04`.

Fetching data using the Fetch API in React

Fetch API is an inbuilt API in a web browser that provides the JavaScript interface for communication over the internet using HTTP. Every web browser has a JavaScript engine as a runtime to compile and run JavaScript code.

The React ecosystem relies inarguably on JavaScript. This is a fact and one of the reasons you are expected to understand modern JavaScript before delving into React application development.

As a React developer, you will need network resources to build web applications. The `fetch()` method provides you with the means to access and manipulate HTTP object requests and HTTP protocol responses. Let's say that in our web application, we want to display the list of conference speakers and their associated data. This information is housed in another resource database server.

From a third-party public API, we are going to consume the user's resource to fetch hypothetical data to be used in our React application as follows:

```
import React, { useEffect, useState } from 'react';
const App = () => {
  const [data, setData] = useState([]);
  const getSpeakers = ()=>{
    fetch("https://jsonplaceholder.typicode.com/users")
        .then((response) => response.json())
        .then((data) => {
          setData( data);
        })

  }
  useEffect(() => {
    getSpeakers()
  },[]);
  return (
    <>
      <h1>Displaying Speakers Information</h1>
      <ul>
```

```
        {data.map(speaker => (
          <li key={speaker.id}>
            {speaker.name},  <em> {speaker.email} </em>
          </li>
        ))}
      </ul>
    </>
  );
};


export default App;
```

Let's discuss the preceding `fetch` data snippet in detail:

- `import React, { useEffect, useState } from 'react'`: This line imports React's core functions and some Hooks for use in our component.

- Initializing the `useState`: We initialize our state by calling `useState` in our component as follows:

  ```
  const [data, setData] = useState([]);//using a destructuring
  array to write concise code.
  ```

  The `useState` accepts an initial state of an empty array (`useState([])`) and returns two values, `data` and `setData`:

  - `data`: The current state

  - `setData`: State updater function (this function is responsible for the new state of the initial state)

- `useState([])` is `useState` with the initial value of empty array `[]`

- The following snippet has a global `fetch()` method that accepts the URL of the endpoint, `https://jsonplaceholder.typicode.com/users`, where we have the hypothetical resource for speakers:

  ```
  const getSpeakers = ()=>{
    fetch("https://jsonplaceholder.typicode.com/users")
      .then((response) => response.json())
      .then((data) => {
        setData( data);
      })
  ```

  The URL in the preceding code is our resource endpoint. It returns JSON as data. `setData()` accepts the new state, which is the returned data in JSON.

- The `useEffect` Hook is used to invoke the `getSpeaker` function:

```
useEffect(() => {
getSpeakers()
    }, []);
```

- `map()` is invoked on the data array and used to iterate on the speakers' data to display details on the screen:

```
{data.map(speaker => (
    <li key={speaker.id}>
        {speaker.name},  <em> {speaker.email} </em>
    </li>
    ))}
```

In sum, `fetch()` accepts the resource URL (`https://jsonplaceholder.typicode.com/users`) as an argument, which is the path to the resource over the network we are interested in, and returns to a promise that moves to the status of fulfilled once the requested resource response is available.

> **Note**
>
> In real-world applications, it's crucial to manage network errors effectively, especially when data retrieval encounters issues or when the absence of data occurs. Additionally, implementing a loading state can significantly enhance the overall user experience.

Next, we are going to look at another technique for fetching data in React projects using `async/await` using the ECMAScript 2017 feature.

Fetching data using async/await syntax

There are three ways to write asynchronous codes in vanilla JavaScript: callbacks, promises, and `async/await`. In this section, we are going to focus on `async /await` and see how it can be used in React web applications. `async/await` is an improvement on promises.

The following snippet explains how `async/await` can be used to fetch data from an API using a promise-based approach:

```
import React, { useEffect, useState } from 'react';

const App = () => {
    const [data, setData] = useState([]);
    const API_URL = "https://dummyjson.com/users";

    const fetchSpeakers = async () => {
        try {
```

```
                const response = await fetch(API_URL);
                const data = await response.json();
                setData(data.users);
            } catch (error) {
                console.log("error", error);
            }
        };


    useEffect(() => {
        fetchSpeakers();
    }, []);


    return (
      <> [Text Wrapping Break]

        <h1>Displaying Speakers Information</h1>
[Text Wrapping Break]

        <ul>
            {data.map(item => (
                <li key={item.id}>
                    {item.firstName} {item.lastName}
                </li>
            ))}
        </ul>


    </>
    );
};

export default App;
```

Let's discuss the preceding code snippet, which demonstrates how to use `async/await` to fetch data asynchronously:

- `import React, { useEffect, useState } from 'react'`: This line imports React core functions and some Hooks for use in our component.

- Initializing the `useState`: We initialize our state by calling `useState` in our component as follows:

  ```
  const [data, setData] = useState([]);//using a destructuring
  array to write a concise code.
  ```

useState accepts an initial state of empty array (useState([])) and returns two values, data and setData:

- data: The current state

- setData: State updater function (this function is responsible for the new state of the initial state)

- useState([]) is useState with the initial value of empty array [].

  The following snippet has a global fetch() method that accepts the URL of the endpoint (https://dummyjson.com/users) and asynchronously returns data once it is available:

  ```
  const API_URL = "https://dummyjson.com/users";

  const fetchSpeakers = async () => {
      try {
          const response = await fetch(API_URL);
          const data = await response.json();
          setData(data.users);
      } catch (error) {
          console.log("error", error);
      }
  };
  ```

  We have the hypothetical resource for speakers in the preceding endpoint. This is our resource endpoint. It returns JSON as data. setData() accepts the new state, which is the returned data in JSON.

- The useEffect Hook is used to invoke the fetchSpeakers function, which fetches data asynchronously from the endpoint const API_URL = "https://dummyjson. com/users":

  ```
  useEffect(() => {

      fetchSpeakers();
  }, [data]);
  ```

  The array dependency is supplied with the data state. When the data state changes, maybe because of the addition or removal of speakers in the list, the component re-renders and shows the updated state.

- Finally, map() is invoked on the data, and it is used to iterate on the speakers' data to render details to the screen:

  ```
  return (
          <>
  ```

```
            <ul>
            {data.map(item => (
                <li key={item.id}>
                    {item.firstName} {item.lastName}
                </li>
            ))}
            </ul>
```

The `async`/`await` approach of fetching data gives your codes better organization and improves the responsiveness and performance of your React applications. The non-blocking mode of `async`/ `await` means you can carry on with the rest of the code operations while you await responses from a large amount of data running tasks.

Next, we are going to look at another approach for fetching data from an API, using a third-party npm package called Axios.

Fetching data using Axios

Axios is simply a lightweight JavaScript, promise-based HTTP client used to consume API services. It is mainly used in the browser and Node.js. To use Axios in our project, open the project terminal and type the following:

```
npm install axios
```

Now, let's see how to use Axios in the following code snippet:

```
import React, { useEffect, useState } from 'react';
import axios from 'axios';

const App = () => {
    const [data, setData] = useState([]);
    const getSpeakers = ()=>{
        axios.get(
            "https://jsonplaceholder.typicode.com/users")
            .then(response => {
                setData(response.data)
            })
    }
    useEffect(() => {
        getSpeakers()
    }, []);
    return (
        <>
```

```
        <h1>Displaying Speakers Information</h1>

        <ul>
            {data.map(speaker => (
                <li key={speaker.id}>
                    {speaker.name},  <em>
                        {speaker.email} </em>
                </li>
            ))}
        </ul>
    </>
  );
};

export default App;
```

Let's examine the preceding code snippet to see how Axios can be used in data fetching:

- `import React, { useEffect, useState } from 'react'`: This line imports React core functions and some Hooks for use in our component.

- `import axios from "axios"`: This line brings in the already installed Axios package for use in the project.

- Initializing `useState`: We initialize our state by calling `useState` in our component as follows:

  ```
  const [data, setData] = useState([]);//using a destructuring
  array to write a concise code.
  ```

- `useState` accepts an initial state of an empty array (`useState([])`) and returns two values, `data` and `setData`:

  - `data`: The current state

  - `setData`: State updater function (this function is responsible for the new state of the initial state)

- `useState([])` is `useState` with the initial value of empty array `[]`.

  The following snippet has a global `fetch()` method that accepts the URL of the endpoint `https://jsonplaceholder.typicode.com/users`:

  ```
  const getSpeakers = ()=>{
      axios.get(
          "https://jsonplaceholder.typicode.com/users")
          .then(response => {
              setData(response.data)
  ```

```
        })

    }
```

This is our resource endpoint. It returns JSON as data. The `setData()` accepts the new state, which is the returned data in JSON.

The `getSpeakers` function uses `axios.get()` to fetch external data from the endpoint and return a promise. The state value is updated, and we have a new state in `setData` from the response object:

```
useEffect(() => {
getSpeakers()
    }, []);
```

* The `useEffect` Hook is used to call `getSpeaker()` and renders the component:

```
<ul>
    {data.map(speaker => (
        <li key={speaker.id}>
            {speaker.name},   <em> {speaker.email}
                </em>
        </li>
    ))}
</ul>
```

Finally, `map()` is used to iterate on the speakers' data and display names and emails on the screen.

Moving on to data fetching techniques in React, we are going to look at another approach to fetching data using React Query.

Fetching data using the React Query in React

React Query is an npm package library created for data fetching purposes with a ton of functionalities loaded with it. In React Query, the state management, pre-fetching of data, request retries, and caching are handled out of the box. React Query is a critical component of the React ecosystem with over a million downloads weekly.

Let's refactor the code snippet we used in the *Fetching data using Axios* section and experience the awesomeness of React Query:

1. First, install React Query. In the root directory of the project, do the following:

```
npm install react-query
```

2. Inside App.js, add the following:

```
import {useQuery} from 'react-query'
import axios from 'axios';
function App() {
  const{data, isLoading, error} = useQuery(
    "speakers",
    ()=>{ axios(
      "https://jsonplaceholder.typicode.com/users")
  );
  if(error) return <h4>Error: {error.message},
    retry again</h4>
  if(isLoading) return <h4>...Loading data</h4>
  console.log(data);
  return (
      <>
        <h1>Displaying Speakers Information</h1>
        <ul>
          {data.data.map(speaker => (
            <li key={speaker.id}>
              {speaker.name},  <em>
                {speaker.email} </em>
            </li>
          ))}
        </ul>
      </>
  );
}
export default App;
```

Check the preceding *Fetching data using Axios* section to compare the code snippet. The React Query snippet is way shorter and more concise. The need for useState and useEffect Hooks have been handled out of the box by the useQuery() Hook.

Let's dissect the preceding code:

- useQuery accepts two arguments: the query key (speakers) and a callback function that uses axios() to fetch a hypothetical speaker from the resource endpoint.
- useQuery is destructured with variables – {data, isLoading, error}. We then check to see whether there is an error message coming from the error object.
- Once we have data, then the return() function returns an array of speakers' data.

Inside `index.js`, add the following code. The existing `index.js` codes are presumed present:

```
import { QueryClient, QueryClientProvider } from
    react-query";
const queryClient = new QueryClient();
root.render(

    <QueryClientProvider client={queryClient}>
        <App /> </QueryClientProvider>
);
```

Let's have some explanation of the code snippet in `index.js`:

- Import the `{ QueryClient, QueryClientProvider }` from React Query: `QueryClient` allows us to leverage the global defaults for all queries and mutations in React Query. The `QueryClientProvider` connects and provides a `QueryClient` to the application.

- Create a new `QueryClient` instance, `queryClient`: Wrap your component with `QueryClientProvider`—in this case, `<App />` is the component—and pass the new instance as an attribute value.

Now run `npm start` if `localhost:3000` is not running. The following should be displayed on the screen:

← → C ⓘ localhost:3000

Displaying Speakers' Names and Emails

- Leanne Graham, *Sincere@april.biz*
- Ervin Howell, *Shanna@melissa.tv*
- Clementine Bauch, *Nathan@yesenia.net*
- Patricia Lebsack, *Julianne.OConner@kory.org*
- Chelsey Dietrich, *Lucio_Hettinger@annie.ca*
- Mrs. Dennis Schulist, *Karley_Dach@jasper.info*
- Kurtis Weissnat, *Telly.Hoeger@billy.biz*
- Nicholas Runolfsdottir V, *Sherwood@rosamond.me*
- Glenna Reichert, *Chaim_McDermott@dana.io*
- Clementina DuBuque, *Rey.Padberg@karina.biz*

Figure 4.1 – Screenshot showing the usage of React Query in fetching data

React Query is very effective at fetching data from API resources. It encapsulates functions that may be required by useState and useEffect. React Query radically redefines the way we fetch data in React applications by introducing a powerful caching mechanism with the queryKey.

Instead of manually managing data fetching and caching, React Query handles it transparently. React Query allows developers to easily fetch and cache data with just a few lines of code, reducing boilerplate and improving performance.

The library provides various Hooks and utilities that simplify data fetching, error handling, and data synchronization with the server, leading to a more efficient and seamless user experience. Exploring React Query further can open up a world of possibilities in handling complex data fetching scenarios and optimizing data management in React applications.

Summary

Handling data is a critical component of any web application. React has proven to be very efficient and scalable in handling data at scale. In this chapter, we discussed various approaches you can utilize in your project to handle data fetching. We discussed fetching data using the Fetch API, async/await, Axios, and React Query.

In the next chapter, we are going to discuss JSX and how you can display lists in React.

5
JSX and Displaying Lists in React

Componentization is a design paradigm in React application development. As a developer and React enthusiast, you will develop tons of useful components. You will need a combination of units to provide interfaces the user can interact with seamlessly.

JavaScript Syntax Extension (JSX) is an innovative approach to describing the **User Interface** (UI) for modern web applications. In this chapter, we are going to take a clinical dive into why JSX is one of the core requirements in developing production-ready React applications. In addition, you will learn how to display lists in React.

We use lists in virtually every web application development project we undertake, and knowing how to render lists is a required skill set for web developers. HTML and JavaScript, as languages of the web, have been with us from the beginning, helping web developers build web applications.

However, in recent times, the demand for complex and highly rich interactive web applications has necessitated using JSX as a creative approach for building user interface components.

In this chapter, we are going to understand what JSX is about and how it is different from HTML. We will use JSX to describe the user interfaces we will be building in this chapter. Then, we will examine how we handle event operations in React.

As a React developer, you will consume both internal and external API data for your user consumption. By the end of this chapter, you will be able to display list objects to your users, handle common events in React, and render lists with loop functions.

In this chapter, we'll cover the following topics:

- What is JSX?
- JSX versus HTML
- How JSX abstracts JavaScript

- Event handling in React

- Displaying lists in React

- Nesting lists in JSX

- Looping over objects in JSX

Technical requirements

The complete code for this chapter is available on GitHub at: `https://github.com/PacktPublishing/Full-Stack-Flask-and-React/tree/main/Chapter05`.

What is JSX?

You've already been introduced to and seen some JSX. Let's discuss in more depth what JSX means as a new approach to adding HTML to JavaScript when designing user interfaces.

JSX is simply an XML-like syntax extension for JavaScript. JSX allows frontend developers to bake HTML elements with JavaScript. The effect of this mix is usually an impressive user-friendly interface. As we know, the main purpose of React is to provide us with a set of APIs for building user interfaces.

With little or no controversy, React has been up to the challenge, becoming the leading shining gem in the jungle of frontend JavaScript libraries and frameworks. React powers large-scale, production-grade web and mobile applications with an improved user experience.

Interestingly, React is achieving this improved efficiency and performance with the same set of tools, languages, and techniques we are already familiar with: HTML and JavaScript. React leverages HTML elements and JavaScript functions to build reusable UI components. JSX evolved as an approach that allows us to mix markup and display logic for building React components.

You can safely run JavaScript code as a JSX expression. Consider the following snippet of a simple React component to see some of the ways you can use JSX in React:

```
import React from 'react';
export function App() {
    const speakerName = "John Holt"
    return (
        <div className='App'>
            <h2>{speakerName}</h2>/* This outputs  John
                Holt */
            <h2>{5 + 5 }</h2>/* This outputs the sum of 5 +
                5 = 10 */
        </div>
    );
}
```

Let's examine what is going on with this code snippet.

The preceding code explains how to use JSX in React:

- `import React from 'react'` is used to bring React into scope
- `export function App()` describes a function component named `App()` that can be accessed by other components
- The `const speakerName` variable is declared and assigned a value of `John Holt`
- The following part of the preceding code snippet depicts the JSX part of the `component App()` code:

```
return (
    <div className='App'>
        <h1>Displaying name of a conference
            speaker:</h1>
        <h2>{speakerName}</h2>    /* This outputs John
                                        Holt */
        <h2>{5 + 5 }</h2>   /* This outputs number 10
                        */
    </div>);
```

The preceding code is a mix of HTML elements (`h2` and `div`) and a JavaScript expression inside curly braces (`{speakerName}`). This displays the text `John Holt` while `{5 + 5 }` displays the result of the sum of 5 + 5.

Whatever JavaScript expression you have can be put inside a curly brace in JSX and it will return the expected valid JavaScript output. However, the browser doesn't know what to do with JSX syntax by default; but with the help of the Babel compiler, JSX code is transformed into equivalent JavaScript syntax that the browser natively understands.

The JSX transpilation by Babel contributes significantly to factors that make React applications so brazenly fast. It doesn't just transpile JSX codes into browser JavaScript; it optimizes as well.

You can also see how the `<div className='App'>` attribute is used in JSX; the naming convention for the class attribute is important. We write it in camelCase format – `className` in React. The `className` attribute is assigned a value of `App`, which is used in the CSS file to add style to the component.

In addition, we need to understand that a high-level connection exists between JSX and the **Document Object Model (DOM)**. The DOM is an object-oriented representation of a web document. It is a set of APIs that's used to manipulate web documents that can be loaded on a web browser. A typical web application page represents a web document that DOM APIs use to maintain the DOM structure and content.

The DOM manipulation is usually done by JavaScript – a scripting language. You can use JavaScript objects to create, update, and remove HTML elements. The DOM manipulation is the bedrock of interactivity you see with most web applications. But React handles DOM differently and at best with some creativity.

The React team has been able to identify the challenges with DOM tree re-rendering on every HTML element operation (create, update, and delete), and decided to develop a **virtual DOM** (**VDOM**). VDOM is an abstraction of the native browser DOM that enables React applications to be fast and efficient and exhibit cross-browser compatibility.

React components only re-render the changed node (h1, div, and p – all these represent nodes on HTML) of a DOM, rather than causing the entire web document to be re-rendered on a single node change.

Next, we will discuss how JSX and HTML elements are used to design UI components and the inherent differences between JSX and HMTL.

JSX versus HTML

JSX is an XML-like syntax extension for JavaScript. It is a creative approach to writing HTML inside JavaScript. Technically, React allows us to use what we love: HTML tags in marking up user interfaces while it uses React.createElement() under the hood. JSX makes component interface development hassle-free while optimizing efficiency.

HTML is the standard language for structuring the web. HTML elements power every web page you see on the internet. HTML syntax is easy to understand, and it is the language the browser understands natively.

The following table clearly states the subtle differences that exist between JSX and HTML for better understanding and usage in React applications:

	HTML	JSX
Native to the browser	HTML elements are native to the browser.	JSX is transpiled into JavaScript using Babel before browsers can understand its syntax.
Attribute usage	You have flexibility regarding how you name your HTML attributes, though this is mostly in lowercase, such as onmouseover, onclick, onsubmit, onload, onfocus, and so on.	You must follow the camelCase rule in naming attributes in JSX and event references such as onClick, onChange, onMouseOver, and so on.

	HTML	JSX
The naming of the `class` and `for` attributes	You must use lowercase `class` when naming CSS classes and `for` when naming input labels in HTML.	In JSX, you must use `className` (camelCase) and `htmlFor` for input labels.
Handles JavaScript code	You must use the `<script>...</script>` script tag or an external JS file to add JavaScript to HTML.	In JSX, you can write JS expressions inside curly braces; for instance, `{ new Date().toString }`.

Returns a single parent element	In HTML, you are permitted to return HTML elements without enclosing them in a single parent; for example:
	
	`<div > </div>`
	
	`<p>...</p>`
	
	`<ul>...</ul>`
	
	`<span>...</span>`.
	
	All these tags can independently stay on a web page with enclosing tags.
		In JSX, you must return a single parent element; otherwise, you will get JSX errors; for instance:
		
		`<div></div>` or a fragment tag, `<> </>`, must enclose all your HTML elements:
		
		`<div>`
		
		`<p>...</p>`
		
		`<ul>...</ul>`
		
		`</div`
		
		Or
		
		`<>`
		
		`<p>...</p>`
		
		`<ul>...</ul>`
		
		`</>`

Self-closing tags	In HTML, you can have a self-closing tag without a forward slash; for example, ` `.	In JSX, you must add a forward slash to any self-closing tag; for example, ` `.

Table 5.1 – Differences between JSX and HTML

JSX and HTML allow you to structure web content and enable users to interact with web application interfaces. As a React developer, you must be conversant with the inherent differences between HTML and JSX elements to avoid being red-flagged by the JSX compiler.

Next, we will discuss how JSX allows us to describe a UI with HTML-like tags while it leverages the power of JavaScript under the hood.

How JSX abstracts JavaScript

Nowadays, coding React applications without JSX is not recommended, though it is possible. For instance, you can write a `React.createElement(component, props, ...children)` function to describe a UI.

However, you can easily describe a button UI in JSX with the following code:

```
<Button color="wine">
    Click a Wine Button
</Button>
```

Writing the preceding code without JSX would require you to describe a button UI with the following code:

```
React.createElement(
Button,
    {color: 'wine'},
    ' Click a Wine Button')
```

Doing this in a large React project could lead to multiple issues, such as having to deal with more bugs in your code base and facing a steeper learning curve to become a code-savvy developer who could function optimally at writing this low-level code to describe a UI. However, with very little to disagree on, you would agree that JSX is a better route to toll rather than plain React-modified JavaScript at describing a UI component.

Let's examine how JSX abstracts JavaScript in its data presentation by providing a syntactic creamy chocolate on top of a low-level React function: `React.createElement()`. This implies how React takes on the burden of converting JSX into JavaScript using Babel to make seamless DOM interaction possible.

In `src/index.js`, update the file with the following snippet to see how you can write `React Conference 2024` to a screen without JSX:

```
import React from 'react';
import ReactDOM from 'react-dom/client';
const root = ReactDOM.createRoot(document.getElementById('root'));
root.render(React.createElement('div', {}, 'React Conference 2024'));
```

In the preceding code snippet, `React.createElement()` is a function call with three parameters: `div`, `{ }`, and the expected output text of `React Conference 2024`:

- The `div` parameter in the `React.createElement` function represents the type of HTML element we are creating. This could be any HTML element or tags (`h1`, `p`, `ul`, `li`, and so on). You could even add components as your first parameter.

- The empty curly brace parameter, `{ }`, represents props. It could be an object or null.

- The third parameter represents what we want to see on the screen. This could be ordinary text or a child component.

In `src/app.js`, update the file with the following snippet that explains the use of JSX:

```
import React from 'react';
export function App() {
    return (
        <div className='App'>
            <h1>React Conference 2024</h1>
        </div>
    );
}
```

The preceding snippet shows the JSX constructs of the code that displays `React Conference 2024` – that is, `<div className='App'>`, `<h1>React Conference 2024</h1>`, and `</div>`.

While these look like regular HTML elements, they are JSX expressions. Now, reset `index.js` as follows to describe a UI with JSX:

```
import React from 'react';
import ReactDOM from 'react-dom/client';
import { App } from './App.jsx'
ReactDOM.createRoot(document.querySelector('#root'))
    .render(<App />)
```

Run the code with `npm start`; you will see the text **React Conference 2024** on your browser screen:

React Conference 2022

Figure 5.1 – Screenshot showing the JSX output

To summarize, JSX is a useful tool within the React community that allows developers to meet the presentational needs of components without a hardcore approach to DOM manipulation. It facilitates a smooth UI experience and rich interactivity for end users. With this, we have a fast, efficient, and independent wide range of browser-compatible web applications.

Next, we will discuss event handling in React. Traditional event methods such as onclick, onfocus, onblur, and others are closely related to what we have in React with some subtle naming differences.

We are going to discuss this and more to see how we can harness the knowledge covered in this chapter and add it to the skill set required to build an interactive conference web app project for this book.

Event handling in React

React's event system is another powerful feature shipped with React core APIs. It is called SyntheticEvent. As React developers, we will come across event handling daily in React application development projects. Handling events shouldn't be new to you if you are familiar with the basics of JavaScript. You could add an event to HTML DOM using the browser-native approach.

Let's have a glimpse at this code snippet:

```
<html>
<body>
<h1>HTML DOM Operations</h1>
<p><strong>Click here to see my message.</strong></p>
<div id="root"></div>
<script>
document.addEventListener("click", function(){
document.getElementById("root").innerHTML =
    "This is a text added to the DOM tree!";
});
</script>
</body>
</html>
```

<div id="root"> </div> indicates the location where the DOM will inject the text we will create. The div element has an id attribute with a value of root passed into it. This helps the event objects know where text should appear. The document.addEventListener() method adds an event listener with two parameters: click and a callback function.

When we click on a button, we fire an event. The event name is called click event. In this scenario, there is a Post message to click to fire an event: <p><strong>Click here to see my message.</strong></p>. Once this message is clicked, a callback function, the second parameter, is triggered and causes the getElementById window method to use the id attribute with a value of root from the div element to pass innerHTML the newly assigned text – that is, This text is then added to the DOM tree!

In React, we have various events with a camelCase naming convention: onClick, onChange, onInput, onInvalid, onReset, onSubmit, onFocus, onBlur, onToggle, and others. You can find a whole list of events in the React documentation – **React SythenticEvents** (https://reactjs.org/docs/events.html)

Let's dive into the following snippet to see how we can handle form events in React. This snippet illustrates the onChange form event:

```
import React,{useState} from 'react';
const App = ()=> {
const [username,setUsername]= useState("");
const [name,setName]=useState("");
const [email,setEmail]=useState("");
const handleSubmit=(e)=>{
    e.preventDefault()
    alert(`Username:${username}, Name: ${name} and Email:
        ${email} submitted`)
  }
    return (
        <div>
            <form onSubmit={handleSubmit}>
                <label htmlFor="username"> Username</label>
                    <br />
                <input type="text" placeholder="Username"
                    onChange={ (e)=>setUsername(
                        e.target.value)} /><br />
                <label htmlFor="name">Name</label><br />
                <input type="text" placeholder="Name"
                    onChange={ (e)=>setName(e.target.value)}
                        /><br />
                <label htmlFor="email"> Email</label><br />
                <input type="email" placeholder="Email"
                    onChange={ (e)=>setEmail(
                        e.target.value)}/><br />
                <button>Submit</button>
            </form>
            <div>
                <p>Username: {username}</p>
                <p>Name: {name}</p>
                <p>Email: {email}</p>
            </div>
        </div>
    );
}
export default App;
```

Let's closely examine this snippet:

- `import React, {useState} from 'react'` makes the `useState` hook available to the `App()` component for state tracking.

- The `const [username,setUsername]=useState("")` `const name, setName]=useState("")` and `const [email,setEmail]=useState("")` snippets allow us to have access to the states for username, email, and name. With `setUsername`, `setName`, and `setEmail`, we can track the new states for the variables. The initial states for all the variables are set to empty strings.

  Now, let's look at the following part of the snippet:

  ```
  const handleSubmit=(e)=>{  e.preventDefault()
      alert(`Username:${username}, Name: ${name} and
          Email: ${email} submitted`)
  }
  ```

  In the preceding snippet, `handleSubmit()` is an event handler that accepts e as the event object. `e.preventDefault` prevents browser default behavior on form field submission. There won't be a page refresh on form submission. `handleSubmit` is added to the form as an attribute to execute `alert()` to display the `username`, `name`, and `email` states on the screen.

- In the input tags, `<input type="text" placeholder="Username"onChange= {(e)=>setUsername(e.target.value)} />`, the onChange event attribute is added to the input element. The onChange event has a callback function that accepts e as a parameter. `setUsername(e.target.values)` listens for a change event in the input field. The same onChange event callbacks are added to input tags for the name and email.

- The `<p>Username: {username}</p>`, `<p>Name: {name}</p>`, `<p>Email: {email}</p>` snippet displays the current input text in the input form fields.

  The following screenshot shows the effect of event handling in forms:

Figure 5.2 – Screenshot showing the effects of event handlers on form fields

To summarize, event handling in React is managed by a cross-browser wrapper called `SyntheticEvent`. We pass event handlers as instances to `SyntheticEvent` in React applications. It is best practice to prevent the default browser behavior by using `preventDefault()`.

Next, we will discuss how to use lists in React applications. A list is a common feature in any enterprise web application. Let's see how React helps us improve the user experience with a well-thought-out UI list design.

Displaying lists in React

Most web applications we see around today use list components in describing a user interface. In any complex web application project or production-grade application, you will see the list feature, often used in data presentation. In React, you can use a list to display your component data.

We are going to use mocked data to showcase how you can use `map()` to fetch a list of data items. We will also discuss the essence of the `key` and `id` attributes in React list management. The GitHub repository for this book's project (`https://github.com/PacktPublishing/Full-Stack-Flask-Web-Development-with-React/tree/main/Chapter-05/06/frontend`) contains the mocked conference speakers `data` source; you can find `images` in the public folders and `css` (`index.css`) inside the `src` folder.

However, the backend section of this book (*Chapter 9, API Development and Documentation*) will explain how we can pull this data from an API endpoint developed in Flask.

This `url` leads you to the speakers page – `http://localhost:3000/speakers`:

Figure 5.3 – Screenshot showing the list of conference speakers pulled from a data source

Let's examine the code snippet for showing a list of speakers working with a couple of components:

Inside `src/components`, create `SpeakersCard/SpeakersCard.jsx` and add the following code snippet:

```
import React from 'react'
const SpeakersCard= ({name, jobTitle, company, profileImg}) => {
    return (
        <>
            <div className="card">
                <div className="speaker-card">
                    <div className="speaker-info">
                        <img src={profileImg} alt={name} />
                        <span>
                            <h3>{name}</h3>
                        </span>
                        <p>{jobTitle}</p>
                        <p>{company}</p>
                    </div>
                </div>
            </div>
        </>)
}
export default SpeakersCard;
```

In the preceding snippet, we created a `SpeakersCard` component that accepts an object with four properties: name, `jobTitle`, company, and `profileImg`. These properties will be passed to the component as props by the `SpeakersPage` component (parent component) that will be created shortly.

The `SpeakersCard` component's return statement contains JSX, which represents the structure of the rendered output.

We need this `SpeakersCard` component to encapsulate and represent the visual appearance and information display for a speaker entity for the conference web app. By creating this component, we can reuse it throughout the application whenever we need to display information about the speaker.

Now, inside `src/pages`, create `SpeakersPage/SpeakersPage.js`. The `SpeakersPage` component will be used to display a list of speakers by rendering the `SpeakersCard` component for each speaker in `speakerList`.

Add the following code to `SpeakersPage.js`:

```
import React from 'react';
import SpeakersCard from
    '../../components/SpeakersCard/SpeakersCard';
```

```
import speakerList from '../../data/SpeakerList';
import Breadcrumb from
    '../../components/Common/Breadcrumb/Breadcrumb'
import Header from '../../components/Header/Header';
import Footer from '../../components/Footer/Footer';

const SpeakersPage = () => {
    return (
        <>
            <Header/>
            <Breadcrumb title={"Speakers"}/>
            <div className="speakers-container">
                <div className="section-heading" >
                    <h1>Meet Our Speakers</h1>
                </div>
                <div className="card">
                    {speakerList.map((speaker) => (
                        <SpeakersCard
                            key={speaker.id}
                            name={speaker.name}
                            jobTitle={speaker.jobTitle}
                            company={speaker.company}
                            profileImg={speaker.profileImg}
                        />
                    ))}
                </div>
            </div>
            <Footer/>
        </>
    )

}
export default SpeakersPage;
```

In the preceding snippet, we import dependencies to make the page functionalities work:

- `import SpeakersCard from '../../components/SpeakersCard/SpeakersCard'`: This code line imports the `SpeakersCard` component from the `SpeakersCard` directory within the components directory. The `../../` notation is used to navigate to the appropriate directory level.

- `import speakerList from '../../data/SpeakerList'`: This code line imports the `speakerList` data from the `SpeakerList.js` file located in the data directory. You can find the data file in the GitHub repository for this chapter. This data contains an array of objects, each representing a speaker with properties such as `name`, `jobTitle`, `company`, and `profileImg`.

Then, we add `<SpeakersCard ... />`. This code line renders the `SpeakersCard` component and passes the necessary props (name, `jobTitle`, `company`, and `profileImg`) for each speaker from `speakerList`. The `key={speaker.id}` prop is added to each `SpeakersCard` component. The `key` prop helps React efficiently update and re-render components when the list changes.

`SpeakersPage` also includes `header`, `breadcrumb navigation`, and `footer` components to provide a complete layout for the speakers' section. The codes for the `Header`, `Breadcrumb`, and `Footer` components can be found in this chapter's GitHub repository.

Next, we will examine how we can follow React best practices in dealing with list items by uniquely identifying the items in a list with a key.

Using key and id in JSX

The **key** in React list items is a unique identifier of the state of items in a list. We use a key to track items in the list that have been changed, added, or removed. It is usually expected to be a unique item in a list.

Take a look at the object array we used in the preceding example:

```
const speakerList = [
    {
    id: 1,
    name: 'Advon Hunt',
    jobTitle:'CEO',
    company:'Robel-Corkery',
    profileImg: 'https://images.unsplash.com/photo-
        1500648767791' },
]
```

The `id` attribute in this array should be a unique number. This allows us to track object data states appropriately. We used the `{speaker.id}` ID as the value for the key attribute in the preceding `speakersList` example.

We will now delve into nested lists in JSX and learn how to use a nested list in React to handle complex data structures.

Nesting lists in JSX

As mentioned earlier, lists are a critical component of most web applications. Lists are often used to structure data and organize information neatly. We are familiar with some of the list clichés in web development: *a to-do list*, *a task list*, and even *a menu list*. All these lists can become complicated, depending on the data structure and how you are expected to present list items to end users. Dealing with lists in a React application requires an understanding of how you can handle data that comes in the form of an array of objects.

In this section, we will learn how to render nested lists of items in JSX in React applications. You are going to see complex nested data structure like this and even more coming from your API data sources, so having an understanding of nested lists will make React applications that contain complex data easier to handle.

The following code snippet displays a list of nested web technology stack items in a component.

Edit App.js inside src/App.js:

```
import React from "react";
import {webStacksData} from "./data/webStacksData";
import WebStacks from "./components/WebStacks/WebStacks";
const App = () => {
    return (
        <ul>
            {webStacksData.map(i => (
                <WebStacks item={i} key={i.id} />
            ))}
        </ul>
    );
}
export default App;
```

So, what's happening in this code? We are dealing with a named nested list of object data called webStacksData, which can be found in the GitHub repository for this book:

- The data is imported into scope with import {webStacksData} from "./data/webStacksData";.

- We also imported the WebStacks components into scope.

- The webStacksData.map function iterates over each item in the webStacksData array, creating a new WebStacks component for each item. The key prop is set to the id property of each item to help React efficiently update the list when needed. For each item in the webStacksData array, a WebStacks component is rendered with the item prop set to the current item from the array.

Let's create a component called WebStacks to see the inner working of the component:

```
import React from "react";
const  WebStacks = ({ item })=> {
let children = null;
if (item.values && item.values.length) {
    children = (
        <ul>
            {item.values.map(i => (
```

```
                <WebStacks item={i} key={i.id} />
            ))}
        </ul>
    );
}
return (
    <li>
        {item.name}
        {children}
    </li>
);
}


export default WebStacks;
```

The WebStacks component takes props items. In the component body function, we check to see whether the parent list items exist and whether it has child items. We then invoke map() to recursively iterate over the list items with valid child list items.

This <li>{item.name}{children}</li> returns the name of the list items and all the children. Next, we are going to see how you can loop through objects in React and display the output in JSX.

Looping over objects in JSX

Looping through complex data objects is part of what experienced React developers need to know how to handle effortlessly. You will undoubtedly encounter scenarios where you will have to work with both simple and nested object data from your API endpoints to extract useful data for your application. In this section, we are going to understand how to seamlessly iterate over data objects in an application.

In JavaScript, objects are not iterable. You simply can't loop over the object properties with the for ... of syntax. Object.Keys() is one of the in-built standard object methods used to loop over object data in JavaScript. However, in ES2017, new object methods were added that can be used to loop over object properties: Object.values() and Object.entries().

Let's briefly examine each of these methods and learn how to use them with object data.

Create the object data to loop over and name it speakersData:

```
const speakersData = {
name:"Juliet Abert",
company:"ACME Group",
street:"1st Avenue",
state:"Pretoria",
country:"South Africa"
}
```

Next, we will examine a variety of techniques that are used to efficiently iterate over object properties, allowing you to access and manipulate data within objects using methods such as `Object.keys()`, `Object.values()`, and `Object.entries()`. We will briefly delve into each of these techniques, starting with `Object.keys()`.

Using Object.keys()

The `Object.keys` method returns an array of object's keys. As you know, objects contain key and value pairs, so `Object.keys()` will return arrays of keys/properties.

Let's pass in our data object as a parameter in the following code snippet:

```
console.log(Object.keys(speakersData));
```

We will get the following output:

Figure 5.4 – Screenshot showing the effect of using the Objects.keys() method

Here, you can see the array of keys. You can retrieve the values of the key with a loop function:

```
for (const key in speakersData){
  console.log(`${key}: ${speakersData[key]}`);
}
```

The following screenshot shows the keys and values of the object data:

Figure 5.5 – Screenshot showing object keys and values

Later, you can invoke `map()` to retrieve the values of the keys in a React component. This will be explained later in this section.

Now, let's learn how to use `Object.values()`.

Using Object.values()

The `Object.values()` method returns an array of object property values:

```
console.log(Object.values(speakersData));
```

This returns only the properties values without keys, thus making it less useful in a use case where keys and values are needed.

Figure 5.6 – Screenshot showing the effect of using the Objects.values() method

Let's look at the last technique we can use to loop over object data.

Using Object.entries()

The `Object.entries()` method returns an array of object key-value pairs – `[key, value]`. Looping over an object with `Object.entries()` is easier with the `[key value]` pair. For instance, consider the following code:

```
for (const  key of Object.entries(speakersData) ){
    console.log(`${key[0]} : ${key[1]}`)   }
```

The following screenshot shows the output of using `Object.entries()` on object data:

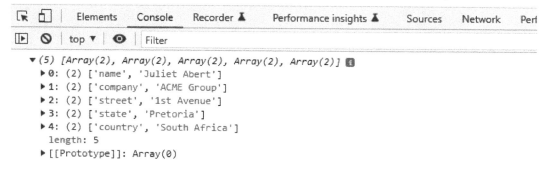

Figure 5.7 – Screenshot showing the effect of using the Objects.entries() method

We can see that 2D arrays are returned with keys and values of object properties.

Example of looping with Object.keys

Now, we are going to work with a data object that contains useful speakers' information with an object data format. It can be found in this book's project repository on GitHub (https://github.com/PacktPublishing/Full-Stack-Flask-Web-Development-with-React/blob/main/Chapter-05/data/objSpeakersData.js) and displays the output in a React component.

The following code iterates over the speakers object data and displays the output in JSX:

```
import React from 'react';
import {simpleSpeakerData} from
    '../../data/objSpeakersData';
const Speakers = () => {
    return (
        <>
            <h1>Speakers</h1>
            <div>
                <ul>
                    {Object.keys(s).map(key => (
                        <li key=
                            {key}>{simpleSpeakerData[key]
                                .name}
                            {simpleSpeakerData[key].company}
                            {simpleSpeakerData[key].street}
                            {simpleSpeakerData[key].state}
                            {simpleSpeakerData[key].country}
                        </li>
                    ))}
                </ul>
            </div>
        </>
    );
}
export default Speakers;
```

The preceding code is explained as follows:

- import {simpleSpeakerData} from '../../data/objSpeakersData' brings our data into scope so that it can be used in the code.

- Then, we declare a Speakers component, which returns a list of object data.

- simpleSpeakerData is passed to Object.keys(simpleSpeakerData).

- map() is then called on the returned keys from Object.keys(). This iterates over the key arrays that are returned.

  We are now able to access the individual key and values of the object.

- {simpleSpeakerData[key].name} points to the name property value of the object data.

The following figure shows the output of iterating over complex object data in React using JSX:

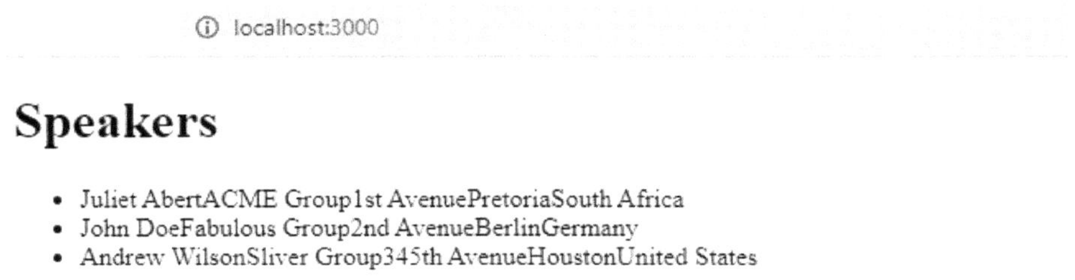

ⓘ localhost:3000

Speakers

- Juliet AbertACME Group1st AvenuePretoriaSouth Africa
- John DoeFabulous Group2nd AvenueBerlinGermany
- Andrew WilsonSliver Group345th AvenueHoustonUnited States

Figure 5.8 – Screenshot showing object data

Looping over objects in JSX using Object.keys(), Object.values(), and Object. entries() is essential and involves iterating through the properties of an object to render JSX elements dynamically. This approach allows you to generate lists, tables, or other UI components that display data from objects in a structured manner.

Summary

In this chapter, we extensively discussed JSX in React. We delved into explaining what JSX is all about as well as the rules guiding the use of JSX in React. Then, we discussed the DOM and how VDOM in React abstracts the native browser DOM for React developers to build a more efficient, cross-browser user interface. JSX improves DOM interaction in React applications and also optimizes the speed for elements in React components to render.

We also examined event handling in React and the use of the SyntheticEvent event wrapper in React in handling event operations in React. We discussed the subtle differences between JSX and HTML and the rules guiding the usage in React.

Finally, with use cases, we discussed how you can display lists in a React project and how key and id are used in managing list items uniquely. We also looked at how you can iterate over objects and display complex nested objects in React.

In the next chapter, we will discuss how to handle form operations in depth and routing in React web applications.

## 6

Working with React Router and Forms

React Router is a library for client- and server-side routing. Imagine the usual way websites work; when you click on a link, your browser sends a request to the web server, receives a bunch of data, and then takes time to process everything before finally displaying the content of the new page.

You will get the same experience every time you request a new page from the website. With client-side routing, things get way smoother! Instead of going through that whole process every time you click a link, your web app can update the URL instantly without bothering the server for a new document. This means your web app can quickly show you a new part of the app without any delays. This and more is what React Router offers.

In this chapter, we will explore React Router v6 as a magical tool to handle navigation. You can also use React Router for data fetching but we will limit our scope to component navigation in this book. You will implement simple and complex nested routes in React. You will also work with the `useParams` and `useNavigate` hooks for dynamic and programmatic routing.

Next, we will delve into form handling in React applications. Forms are vital components in any web application. You can't have a complete soup of web applications without forms. Interestingly, we use forms for a variety of purposes that depend on the business or project requirements.

In React, forms are used in components to allow activities such as user login, registration, search, contact forms, shopping checkout page, event attendees' forms, and a host of others. Forms provide a medium for browser-database server interactions.

We collect data from users of our applications through a form; sometimes, we send users' data to a database or send/save it to other platforms such as email and third-party applications. It all depends on how we intend to handle form data.

In a nutshell, you will learn how to use form elements to facilitate user interaction in your React applications. You will also understand how to leverage React Router, a popular routing library for client-side routing.

By the end of this chapter, you will understand how routing works in React applications by working with the React Router library to navigate your different application endpoints. Finally, you will be able to develop elegant React forms and handle users' information in a React way.

In this chapter, we're going to cover the following main topics:

- Routing with React Router
- Adding React Router in React
- Handling dynamic routes
- Using forms in React
- Controlled and uncontrolled form components
- Handling user input – `Input field`, `TextArea`, and `Select`
- Validating and sanitizing users' data in React

Technical requirements

The code for this chapter can be found at `https://github.com/PacktPublishing/Full-Stack-Flask-and-React/tree/main/Chapter06`.

Due to page constraints, some of the code blocks have been snipped. Please refer to GitHub for the complete code.

Routing with React Router

Routing in a React web application is the ability to navigate seamlessly to and from multiple application components, URLs, pages, and resources, both internal and external. By default, React doesn't include page routing in its library. And as a matter of fact, React's main goal is to allow developers to design the display of the view of a single-page web application.

We all know web applications require multiple views, hence the need for an external library such as React Router to allow for component navigation. Working with a large application would require multiple specialized views. This means we have to solve the problem of navigation left untreated with the React library, but this is where React Router comes in.

React Router is an open source package that's used for component-based routing in React applications. It is popular among React developers and widely used in various React projects. Interestingly, you can use React Router anywhere you intend to run React applications: client side with browsers, on a web server with NodeJS, and even via mobile applications with React Native.

So far, we have kind of taken the *Bizza* application bit by bit, cherry-picking the components and their interaction. Now, we are going to move through the pages of our projects and link them up with React Router.

React Router is composed of some routing features. These features are the nuts and bolts of the inner working of React Router. Knowing them will help us with our understanding of React Router. In the following sections, we will explain some of the commonly used features of React Router.

Let's start with Router, a component that enables navigation and routing within a React application.

Routers

React Router provides different types of routers that allow you to handle routing and navigation in React applications. Each router has its specific use case and benefits. We will briefly discuss some of the commonly used React Router routers:

- `CreateBrowserRouter`: This is a specialized function in React Router v6 that serves as the preferred method for generating a browser router in web projects. By utilizing the DOM History API, it efficiently updates the URL and maintains the history stack. Moreover, it unlocks access to the v6.4 data APIs, encompassing loaders, actions, fetchers, and other React Router functionalities.

- `RouterProvider`: This component in React Router is designed to supply the router instance to all components rendered within its scope. This ensures that the router can be utilized for efficient management of the application's navigation and routing needs.

  The `RouterProvider` component requires a `router` prop as an argument, and this prop serves as the router instance that will be distributed to the components rendered within `RouterProvider`. It is essential to position `RouterProvider` at the highest level in the component tree to ensure that all application components can access the router instance effectively.

- `NativeRouter`: This is an interface that is required to run React Router in React Native, a routing solution for mobile applications. This is outside the scope of this book.

Next, we will discuss components in React Router. Components in React Router allow you to render the user interface for a specific route within a single-page application.

Components

Components in React Router enable you to create a flexible and dynamic routing system within your React applications, making it easier to manage navigation and the state as users interact with your UI. We will briefly discuss some of the commonly used components.

- `Link`: `Link` is a component element that allows users to navigate to another component page upon it being clicked. Under the hood, `react-router-dom` renders a `<Link>` tag to an anchor element, `<a>`, with a real `href` that directs users to the resources it is pointing to.

- NavLink: This works as a `<Link>` tag but with the added feature of indicating an active element in a menu. This is commonly used when you are building a tabbed menu and you want to show which part of the menu is currently selected.

- Route: This is used to render the UI in React Router based on the current location. `Route` has a path and an element as props. This is how it works: whenever the path `Route` component matches the current URL, based on the user click operation, it renders its element. This element could be any component in the application. We will see a live example shortly.

- Routes: This has `Route` as its children. `Routes` works logically in a simple way, like `Route`, except that `Route` or a series of `Route` are children of `Routes`. So, whenever the path of the UI component changes, `Routes` checks all its children `Route` elements to determine the best match for the user request or click path and renders that specific UI.

Next, we will discuss hooks in React Router. Hooks provide a mechanism for interacting with the router's state and executing navigation actions directly within your components. We will discuss hooks such as `useLocation`, `useParams`, and `useNavigate`.

Hooks

React Router offers a range of hooks that empower developers with efficient ways to manage routing, the state, and navigation within their components. We will briefly discuss some of the commonly used hooks:

- useLocation: You can use this hook to perform some side effects whenever you need to track changes in the current location. The `useLocation` hook usually returns the current location object.

- UseParams: You can use this hook to get the parameter from the browser through the current URL matching `<Route path>`. The `useParams` hook returns an object of the key-value pairs of the dynamic params.

- UseNavigate: You can use this hook to programmatically navigate between different routes in your React application without the need for a `history` object or the `Link` component.

Now, it is time to add React Router to our root app and connect our pages.

Adding React Router in React

You need to install React Router to use it in your project. We are going to build the navigation features for the *Bizza* project to connect different components. The navigation tabs will consist of the home, about, speakers, events, sponsors, and contact pages. Let's start coding by entering this command in the project directory's Terminal:

```
npm install react-router-dom@latest
```

Once we have the package installed in the root directory of our project, we can create the home, about, speakers, news, and contact pages components.

Now, we will add content to each of these components:

- Inside `src/pages/HomePage/HomePage.js`, add the following code snippet:

```
import React from 'react';
const HomePage = () => {
    return <div> Home page </div>;
};
export default HomePage;
```

- Inside `src/pages/AboutPage/AboutPage.js`, add the following:

```
import React from 'react';
const AboutPage = () => {
    return <div> About page </div>

}
export default AboutPage;
```

- Inside `src/pages/SpeakersPage/SpeakersPage.js`, add the following:

```
import React from 'react';
const SpeakersPage = () => {
    return <div>Speakers </div>
}
export default SpeakersPage;
```

- Inside `src/pages/EventsPage/EventsPage.js`, add the following:

```
import React from 'react';
const EventsPage = () => {
    return <div>Events page </div>
}
export default EventsPage;
```

- Inside `src/pages/SponsorsPage/SponsorsPage.js`, add the following:

```
import React from 'react'
const SponsorsPage = () => {
    return <div>Sponsors Page</div>
}
export default SponsorsPage
```

- Inside `src/pages/ContactPage/ContactPage.js`, add the following:

```
import React from 'react'
const ContactPage = () => {
    return <div>Contact Page</div>
}
export default ContactPage
```

Now that these components have been set, let's start implementing the React Router functionalities in our application:

1. Inside `src/index.js`, add the following code:

```
import React from 'react';
import { createRoot } from 'react-dom/client';
import {
    createBrowserRouter,
    RouterProvider,

} from 'react-router-dom';
```

The preceding snippet shows the `import` statements required for the client-side routing using React Router:

- `createRoot`: Imports the function to create a root React component for rendering

- `createBrowserRouter` and `RouterProvider`: Import components and functions related to React Router, which provides routing functionality

2. We also need to import all the various components we created earlier. Still inside `index.js`, add the following component imports:

```
import HomePage from './pages/HomePage/HomePage';
import AboutPage from './pages/AboutPage/AboutPage'
import SpeakersPage from './pages/SpeakersPage/SpeakersPage';
import EventsPage from './pages/EventsPage/EventsPage';
import SponsorsPage from './pages/SponsorsPage/SponsorsPage';
import SponsorsPage from './pages/SponsorsPage/SponsorsPage';
```

The preceding imports are the various files and components that will be used in the application.

Please note that all future components that we might want React Router to be aware of can be added as part of the imported files and components of the application. Next, we will set up the routing configuration.

Setting up the routing configuration

In the context of web application development and libraries such as React Router, routing configuration refers to the process of setting up rules or mappings that define how different URLs (or routes) in a web application should be handled. It involves specifying which components or views should be rendered for specific URLs, allowing users to navigate through different parts of the application seamlessly.

With React Router, you can define a list of routes and associate each route with a corresponding component to be displayed when the route is matched. These routes can be static, dynamic (with placeholders), or nested to create a hierarchical structure.

Let's put this to practical use. Add the following code to the index.js file:

```js
const router = createBrowserRouter([
  {
    path: "/",
    element: <HomePage />,
  },
  {
    path: "/about",
    element: <AboutPage/>,
  },
  {
    path: "/speakers",
    element: <SpeakersPage/>,
  },
  {
    path: "/events",
    element: <EventsPage/>,
  },
{
    path: "/sponsors",
    element: <SponsorsPage/>,
  },
{
    path: "/contact",
    element: <ContactPage/>,
  },
    ],
);

createRoot(document.getElementById("root")).render(
  <RouterProvider router={router} />
);
```

The preceding code shows the created `router` object using the `createBrowserRouter` function, which defines the routing configuration for the application. This `router` object sets up different paths and their corresponding React components to be rendered when those paths are matched.

What this means is that when a user navigates to different URLs in the application, the corresponding components will be rendered based on the defined routes; for instance:

- Navigating to `/` will render the `HomePage` component
- Navigating to `/about` will render the `AboutPage` component

Likewise, the rest of the components are rendered and displayed based on the routes and their corresponding components. The `createRoot()` function from the `react-dom` library, `'react-dom/client'`, is used to create a `root` component for intended rendering. It is a newer and more efficient alternative to `ReactDOM.render()`. The `createRoot()` function takes the target DOM element as an argument and returns a `root` component that can be used to render React elements into that target element.

In this case, `createRoot(document.getElementById("root"))` creates a `root` React component that will render its content inside the `<div>` element with the `"root"` ID. In essence, the `createRoot` function is used to create a `Root` object and render the `RouterProvider` component into the `root` DOM element.

The `RouterProvider` component then renders the `HomePage` component, which is the default route for the application. `<RouterProvider router={router} />` uses the `RouterProvider` component from React Router. `RouterProvider` takes a prop called `router`, and the value of this prop is the previously defined router object, which contains the preceding routing configuration. This makes `router` available to the entire application, enabling navigation based on the defined routes.

We'll add links to routes in the next section.

Adding links

Let's improve the navigation menu by adding links to the elements. To add a link to an element, use `<Link to="" >elementName </Link>`. `to=""` allows us to insert the navigation path we intend to go to. Let's see the details of a typical link definition:

```
<nav className="nav">
  <ul>
    <li>
      <Link to="/" className='navlink'>Home</Link>
    </li>
    <li>
      <Link to="/about" className='navlink'>About</Link>
    </li>
```

```
    <li>
      <Link to="/speakers" className='active
        navlink'>Speakers</Link>
    </li>
    <li>
      <Link to="/events" className='navlink'>Events</Link>
    </li>
    <li>
      <Link to="/sponsors" className='navlink'>
        Sponsors</Link>
    </li>
      <li><Link to="/contact" className='navlink'>
        Contact</Link>
    </li>
  </ul>
</nav>
```

Check the GitHub `src/components/Header/Header.jsx` file to learn more about the `Link` definition.

The following screenshot shows `HomePage` with a menu and links:

Figure 6.1 – Screenshot showing routes and links

Next, let's learn how to embed a route into another route so that we have what we call a nested route.

Adding a nested route

Nested routes in React Router provide a structured approach to organizing routes within your application. They facilitate grouping related routes, streamlining navigation between different sections. To implement nested routes, you must utilize the children prop on a Route component.

This prop accepts an array of Route components as its value, defining the child routes that will be rendered when the parent route is matched. For instance, consider the following code snippet, which demonstrates the creation of a nested route for speakers.

Inside `src/index.js`, update the `/speakers` route, as follows:

```
const router = createBrowserRouter([
  {
    path: "/speakers",
    children: [
      {
        index: true,
        element: <SpeakersPage />,
      },
      {
        path: "/speakers/:speakerId",
        element: <SpeakerDetail />
      },
    ],
  },

]);
```

In the preceding code, we have a parent route called `speakers` with the `/speakers` path. The child route for `SpeakerDetail` has the `/speakers/:speakerId` path. The `:speakerId` placeholder in the path is a dynamic parameter that will be replaced with the `:speakerId` value of the speaker when the user navigates to the route.

The `SpeakerDetail` component will be rendered with the detailed information of `speakerId` in the URL. Inside `src/pages`, create `SpeakerDetail/SpeakerDetail.js`; then, add the following code:

```
const SpeakerDetail = () => {
    return (
        <div className='page-wrapper'>
            <h1>This is SpeakerDetail with the ID: </h1>
        </div>
    )
}
export default SpeakerDetail
```

The following screenshot shows the nested route with `http://localhost:3000/speakers/234`:

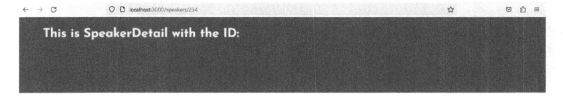

Figure 6.2 – Screenshot showing the nested route

Essentially, nested routes can be used to organize your routes in a way that makes sense for your application. They can also be used to make it easier to navigate between related routes.

Next, we will explore how to handle dynamic and programmatic routes with `useParams` and `useNavigate`.

Handling dynamic routes

In web application development, a **dynamic route** refers to a route that employs a placeholder to represent a changing value. This placeholder can be used to handle various dynamic content. You can have placeholders such as `speakerId`, `productId`, `postId`, and so on to represent the changing value.

For instance, let's consider the preceding speakers route we updated with `/speakers/:speakerId`. It is conventional to add a colon in front of a dynamic route, like so: `:speakerId`. So, how can we retrieve this value of `speakerId` from the URL? This is where the `useParams` hook comes in.

Using useParams

The `useParams` hook in React Router provides access to the dynamic parameters extracted from a route. These parameters are the values that correspond to the placeholders in the path of a dynamic route.

For instance, in the following code snippet, the `useParams` hook is used to retrieve `SpeakerId` from the `/speakers/:speakerId` route. The following code shows the code implementation.

Update the `SpeakerDetail` component in `src/pages/`, as follows:

```
import React from 'react'
import { useParams } from 'react-router-dom'

const SpeakerDetail = () => {
    const {speakerId} = useParams()
    return (
```

```
        <div className='page-wrapper'>
            <h1>This is SpeakerDetail with the ID:
                {speakerId} </h1>
        </div>
    )
}
export default SpeakerDetail
```

In the preceding code snippet, we have the `SpeakerDetail` component, which is used to display the details of a speaker based on the `speakerId` dynamic parameter that was extracted from the URL. The `useParams` hook will return an object that contains the dynamic parameters from the route. In this case, the `speakerId` property of the object will contain the Speaker ID in the URL.

The following screenshot shows the extracted `speakerId` from the URL:

Figure 6.3 – Screenshot showing the extracted speakerId

The `useParams` hook is a powerful tool that can be used to access the dynamic parameters of any route. Next, we will briefly discuss the `useNavigate` hook for programmatic navigation.

Using useNavigate

`useNavigate` is a new hook that was introduced in React Router v6. It provides a way to programmatically navigate or redirect users to different routes within a React application. Unlike the `useHistory` hook from previous versions of React Router that provided access to the history object, `useNavigate` provides a more straightforward and explicit way to navigate between routes.

With `useNavigate`, you can initiate navigation in response to certain events, such as a button click, form submission, or any other user action. Instead of modifying the URL directly like in React Router v5, you can now use the navigate function returned by `useNavigate` to achieve navigation.

For instance, inside the `src/components/Header/Header.jsx` file, we have the following code to show how `useNavigate` is implemented:

```
import React from 'react';
import {Link, useNavigate } from 'react-router-dom';
const Header = () => {
  const navigate = useNavigate();
  const handleLoginButtonClick = () => {
    navigate('/auth/login');
```

```
    }
    return (
      <header className="header">
        ...
        <div className="auth">
          <button onClick={handleLoginButtonClick}
            className="btn">Login</button>

        </div>
      </header>
    );
  }

  export default Header;
```

In the preceding snippet, the useNavigate hook is called to get the navigate function. When the button is clicked, the handleLoginButtonClick function is executed, which, in turn, calls navigate('/auth/login'). This will navigate the user to the '/auth/login' route programmatically.

useNavigate provides a more declarative and concise way to handle navigation compared to directly manipulating the history object. It improves the overall readability and maintainability of the code when working with React Router v6.

This section wraps up routing with React Router. The subsequent section shifts its focus to the realm of managing forms within the React library.

Using forms in React

Conventionally, forms are used to collect user inputs. There is no serious production-grade web application without forms. Using forms in React is slightly different from using HTML form elements. If you have developed React applications for a while, this might not be new to you.

The subtle difference between the elements of React forms and those of normal HTML forms is due to the unique way React handles the internal state of forms. The HTML DOM manages the internal states of native HTML form elements in a browser DOM way. On the other hand, React handles form elements through its components' state.

So, what is this state all about? The state we are talking about is an object that holds user inputs before form submission. Form elements have an internal state that prevents data loss before you submit user input across the processing channel.

Having laid down the background for the internal state management of form elements, let's quickly move on to how React enhances the user experience through its component-based approach, enhanced with the VDOM mechanism of React. We are going to develop forms in React without the use of any external library; instead, we will focus on pure React and leverage its controlled component-based approach in managing form state. Right now, we are going to design a simple sign-up form component.

The following snippet shows a `SignUp` form component to help you understand how to create a simple form in React.

Create `SignUp` inside `src/pages/Auth/SignUp.js/` in your project directory:

```jsx
import React from 'react';

const SignUp = () => {
  return (
    <>
    <div className="signUpContainer">
    <form>
    <h2>Create an account</h2>
      <div className="signUpForm">
        <label htmlFor="name">Name</label>
        <input
          id="name"
          type="text"
          name="name"
        />
      <label htmlFor="email">Email Address</label>
        <input
          id="email"
          type="email"
          name="email"
        />
      <label htmlFor="password">Password</label>
        <input
          id="password"
          type="password"
          name="password"
        />
          <button>Register</button>
      </div>
    </form>
    </div>
  </>
    );
};
export default SignUp;
```

The preceding snippet should look familiar – there's nothing special here except for the `<label>` attribute, `htmlFor`. This is the React way of adding the `for` attributes to the form label. The `htmlFor` prop is used to match the corresponding ID with input form elements.

Moving forward, let's update the router configuration inside `index.js` and add the `signup` route. Inside `src/index.js`, add the sign-up path and its associated component:

```
{
    path: "/auth/signup",
    element: <SignUp/>,
  },
};
```

The following figure shows the output of the signup form code snippet:

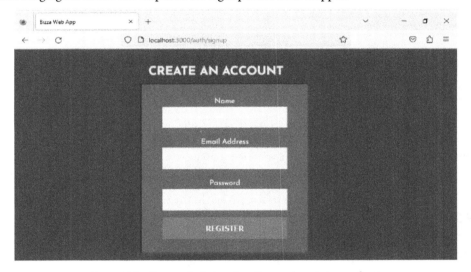

Figure 6.4 – Screenshot showing the rendered signup form

When you navigate to `http://localhost:3000/auth/signup`, you will see that the `SignUp` form component has been rendered. With this, React is just rendering the form elements and allowing native browser DOM to carry on as usual with the page reload at every form submission. If you fill in the form and click **REGISTER**, you will see the page reload effect.

This is certainly an anti-React design pattern, which means this is not a React way of designing form elements. So, what is the React design expectation in building an intuitive user form experience? The answer to this is what we will focus on in the next section. There, you will learn how you can build engaging and reusable forms with all the ingredients of the React sauce.

In React, there are two approaches to form components: controlled and uncontrolled form components. In the next section, we will dive into this and learn how to design form components that enhance smooth interaction with the form elements in a React web application project.

Controlled and uncontrolled form components

So far in this book, we have become familiar with components and how they are the building blocks of any React application. When you blend pieces of independently designed components, you get either a UI component or a full-fledged React web application, depending on what you are working on.

The component-driven approach of React is not going to change anytime soon. Building quality UIs for applications is what React does best. You are going to need a high-performant form one way or the other in your career as a developer, and React has you covered with two approaches to building air-tight form components that prevent data loss and improve user experience regarding form interaction.

These two approaches are controlled and uncontrolled form components. Let's start with controlled form components so that we have a sufficient understanding of how they are implemented and why they are the React-recommended approach to form handling.

Controlled form

In terms of a controlled form, a React component maintains the internal state of user inputs in the form elements. What do we mean? Essentially, React has an in-built event wrapper known as SyntheticEvent, a key component of the React event system.

We spoke a lot about SyntheticEvent in *Chapter 5*, *JSX and Displaying Lists in React*, in the *Event handling in React* section. In the controlled form approach, the event handler function in controlled form components accepts an instance of SyntheticEvent, such as onChange, onInput, onInvalid, onReset, and onSubmit, to control the state of form data.

For instance, the onChange event listens to the change in the state value of the component form: this change could be either a user typing something in the form input or trying to replace the value of the form input. The onChange event is triggered and the state data value is changed appropriately.

Let's explore how event handlers allow us to have controlled form components. The following snippet demonstrates a controlled form in React with the relevant event handlers to manage onChange and onSubmit events. Update the SignUp.js file code to demonstrate the use of event handlers in the form component inside src/pages/Auth/SignUp.js:

```
import React,{ useState } from 'react';
const SignUp = () => {
  const [name, setName ] = useState("");
  const [ email, setEmail ] = useState("" );
  const [password, setPassword] = useState("");
  const nameHandler = (e) => {
    setName(e.target.value);
  };
  const onSubmitHandler = (e) => {
    e.preventDefault();
```

```
      alert(`Name: ${name}: Email: ${email} Password:
        ${password}`);
    };
    return (
      <>
        <div className="signUpContainer">
          <form onSubmit={onSubmitHandler}>
            <h2>Create an account</h2>
            <div className="signUpForm">
              <label htmlFor="name">Name</label>
              <input
                id="name"
                type="text"
                name="name"
                value={name}
                onChange={nameHandler}
              />

              <button>Register</button>
            </div>
          </form>
        </div>
      </>
    );
};
export default SignUp;
```

Please refer to GitHub for the full source code. In the preceding snippet, we have updated our default form by adding some input attributes to it – that is, the `value` and `onChange` events, and set the stage for a controlled form. The form input uses `value` to accept the current value as a prop and `onChange` as a callback to update the state of the value.

As you already know, user actions such as tapping the keyboard, clicking a button on a web page, or mousing over on an HTML element with browsers elicit events. But React has what we call Synthetic Events with a couple of instance methods and properties to listen for user interactions and emit certain events.

Popular among these instances in React are `onChange` and `onClick`. We will use more of these. The `onChange` event is triggered whenever there is a change in the form's input element. The `event.target` property of the Web API is used to access the value of this change.

Furthermore, the `onClick` event gets activated every time an HTML element is clicked; for example, when a button is clicked. In our snippet, event handlers are specified in the body of the component function: `nameHandler`, `emailHandler`, and `passwordHandler`. These event handlers listen to the corresponding value changes in the form inputs and control the operation of the form.

In the case of `nameHandler`, it listens to what the users type in, accesses the value with the use of the `e.target` property, and updates its state with `setName()`:

```
const nameHandler = (e) => {
        setName(e.target.value);
};
```

In the form input with the `Name` label, the following was added:

```
<input
    ...
        value={name}
    onChange={nameHandler}
/>
```

> **Note**
>
> The `input` elements for email and password were equally updated with appropriate values and `onChange` event handlers.

`onSubmitHandler()` handles the `onSubmit` event of the form element:

```
<form onSubmit={onSubmitHandler}>
  const onSubmitHandler = (e) => {
        e.preventDefault();
        alert(`Name: ${name}: Email: ${email} Password:
          ${password}`);
    };
```

`e.preventDefault()` prevents the default reloading behavior of the browser. We also output the submitted form data with `alert()`. What are the benefits of the controlled form component?

There are a few reasons why you would want to use controlled form components:

- React recommends it. It is React's best-practice way of handling user inputs in React.

- The components tightly control the behavior of the form, thereby ensuring a reactive user and developer experience.

- We get instant feedback from forms since the event handlers listen to form elements and emit events appropriately.

The controlled form component improves the experience with form interaction in React applications and it is widely used in the React community. Next, we will learn what the uncontrolled form components provide for the React developer community and how we can use them.

Uncontrolled form

In an uncontrolled form, native DOM maintains and stores the state of user input directly. It does this by storing the values of the form elements in the DOM with a reference to the form elements. This is the conventional way HTML form elements maintain their internal states.

React components simply interact with uncontrolled form elements by maintaining references to the underlying form element in the DOM. Let's replicate our earlier signup form and refactor the snippet to use uncontrolled form components.

This snippet uses the `useRef` hook to reference the values of the form elements in the DOM:

```
import React,{ useRef } from 'react';

const SignUp = () => {
  const onSubmitHandler = (e) => {
    e.preventDefault();
    console.log("Name value: " + name.current.value);
    console.log("Email value: " + email.current.value);
    console.log("Password value: " +
      password.current.value);
  };
  return (     <>
    <div className="signUpContainer">
    <form onSubmit={onSubmitHandler}>
    <h2>Create an account</h2>
      <div className="signUpForm">
        <label htmlFor="name">Name</label>
        <input
          id="name"
          type="text"
          name="name"
          ref={name}

        />
      <button>Register</button>

      </div>
    </form>
    </div>
</>
  );
};
export default SignUp;
```

Please refer to GitHub for the full source code. Let's briefly explain the preceding snippet.

In using the uncontrolled form React component, we needed the `useRef` hook to access the form elements in the DOM. `import React, { useRef } from 'react';` brings React and the `useRef()` hook into scope.

Then, we created a reference variable to hold the reference of the form element in the DOM:

```
const name = useRef();
const email = useRef();
const password = useRef();
```

In the `input` tag, we bound the reference variable to the `ref` attribute of the input element:

```
<label htmlFor="name">Name</label>
        <input
          id="name"
          type="text"
          name="name"
          ref={name}
      />
```

The same steps were taken for the email and password input elements.

To extract the current value of the form fields, we must use the `current.value` property of `useRef` in `onSubmitHandler()`:

```
const onSubmitHandler = (e) => {
  e.preventDefault();
  console.log("Name value: " + name.current.value);
  console.log("Email value: " + email.current.value);
  console.log("Password value: " +
    password.current.value);
};
```

The following figure shows the console logs of the uncontrolled form component in React:

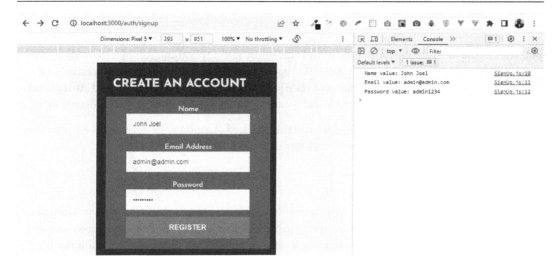

Figure 6.5 – Console log of the uncontrolled form component

What are the benefits of the uncontrolled form component?

Interestingly, there are few benefits to using uncontrolled form components, even though React recommends the controlled approach. Here are a few benefits for consideration in your React project:

- In complex React form applications, where re-rendering the form UI with every user input could be an expensive DOM operation on the performance of the application. Using uncontrolled form components prevents the performance bottlenecks associated with component re-rendering of form elements.

- The uncontrolled form is more appropriate when you need to work with a form input of the `file` type in your React application, such as when doing a file upload.

- The uncontrolled form is quick to use when you are migrating from a legacy non-React code base project. Since DOM maintains the state of the form, working with form fields from the legacy code base is easier.

Now that we have sufficiently examined what an uncontrolled form in React project is all about and covered a few of the benefits of using an uncontrolled form, let's dive into a few of the commonly used input elements: `Input`, `TextArea`, and `Select`.

Handling user input – Input, TextArea, and Select

Handling React form elements is slightly different from the way non-React applications handle user inputs. In this section, we will look at common form elements that are used in handling user input while following React best practices.

Input

Input fields in the form are the most widely used tags in any web application. An input field allows the collection of user data. Input fields have different types depending on their purpose in a form. In the controlled input form element, the component state is always set using a value or checked attribute on the form input field. You also have a callback that listens to the change in value as a result of user input.

With the input type radio and checkbox, we use the checked attributes. To access the value of the input field, we can use `event.target.checked`.

TextArea

`Textarea` is a tag that allows users to write multi-line characters of text. `Textarea` is usually used to collect user data such as comments or review sections in web applications. The `Textarea` element in React works differently. The state is set using a value or checked attribute in the form input field, similar to a single input field. `textarea` doesn't have children, which is typical of HTML.

You can use a callback to retrieve the change in the state value of the form element:

```
<textarea value={textarea} onChange={onChangeCallback} />
```

Select

The `select` form element is used in designing a drop-down list. React has its unique way of working with `select`. The selected value is set with a value attribute on the `select` form element. In React, there is no selected attribute.

This selection is determined by the set value property on the `select` form element. You can use a callback when handling the selected value in the form element:

```
<select value={user} onChange={onChangeCallback}>
<option value="user">User</option>
<option value="admin">Admin</option>
</select>
```

Next, we will discuss how React applications handle user data validation and how you can sanitize your users' data when they're filling in a form element.

Validating and sanitizing users' data in React

Validation is a process that ensures user data quality, integrity, and an appropriate format that's expected for a system. You can never trust data provided by users of your application blindly. While we expect them to trust our code, we can't reciprocate that trust by not guiding them on how our forms and form data should be treated.

Starting as a junior developer, the phrase *Don't ever believe a user would always do the right thing with your form* will forever ring true. You can never trust user data as is. Data that comes from users has to be thoroughly scrutinized and cleaned and ensured it is in the desired format.

Forms fields are the open window into everything you might call the backend in web development. So, trusting user input without some rules in place could be detrimental to your sanity as a developer and to the healthy condition of your web application.

There are always standard-practice validation rules you can adhere to as a React developer. These validation rules guide you and your application against bad actors of your application.

Let's take a look at a few validation rules you might want to check off before user input data is deposited in your database or any of your backend infrastructure:

- **Data type**: You need to ascertain the user is putting in the right data type for a form field. Are they even filling in anything at all? You need to check. For example, in a form field where you are expecting string characters, ensure you are not getting numeric data.

- **Consistency**: You need to ensure data consistency of user input, and one of the ways to be sure you get consistent data is by enforcing validation rules. For instance, you might add a regular expression to check that the length of the password is not less than 8 characters and that it is mixed with symbols.

  Alternatively, you might just allow users to select a country they'd like to visit from a drop-down list of options rather than asking them to type the name of countries they would like to visit. If you do the latter, you will be rudely shocked regarding what you will get in return for your benevolence.

- **Data format**: You probably want the date of birth of your application users to be in *YYYY-MM-DD* or *DD-MM-YYYY* format. You and I know we can't leave this at the discretion of users!

- **Range and constraint**: You may also want to check data against a certain range of parameters or if some data falls within a certain constraint intended. So, you need to enforce this with some mechanisms – a regular expression is an example of such a mechanism. You should always remember that users are prone to making genuine errors as well, even though they don't mean to act badly.

That being said, there are two major instances when you will want to enforce validation rules in React form design:

- **On user input**: As users are interacting with your form elements, you are checking for compliance and giving instant feedback. React shines best here with a controlled form component that uses a callback to harvest user values and relay them to event handlers with the ability to check for errors. We will implement this in our sign-up form shortly.

- **On user submission**: Form data is subjected to validation when the user clicks on the submit button in this instance. This used to be the gold standard in the past. However, nowadays, this seems to be happening less in enterprise application development due to the arrays of frontend technologies that make instant feedback a cool breeze to implement.

Implementing form validation

Let's examine the implementation of form data validation using the React-controlled form component.

We will start by importing the useState and useEffect hooks and bringing them into scope to manage state and side effects, respectively:

```
import React,{ useState,useEffect } from 'react';
```

Then, we must set the state variables as initialValues, formValues, formErrors, and isSubmit:

```
const SignUp = () => {

const initialValues = { name: "", email: "", password: "" };
const [formValues, setFormValues] = useState(initialValues);
const [formErrors, setFormErrors] = useState({});
const [isSubmit, setIsSubmit] = useState(false);
```

initialValues is declared as an object to hold the initial form input state, which is set to empty strings. useState takes the initialValues variable values as the initial state and assigns it to formValues so that from the onset, all the form input values are empty strings. The initial state of formErrors is set to an empty object as well and the initial state of isSubmit is set to false. This means no form has been submitted yet.

We need an onChange function to track changes to form input values. We must set onChangeHandler(), which takes e as a parameter of the event object and destructures the e.target object, which returns two properties – name and value. The setFormValues function accepts all the current formValues with the . . . formValues spread operator and updates them to the new values:

```
const onChangeHandler = (e) => {
const { name, value } = e.target;
setFormValues({ ...formValues, [name]: value });   };
```

useEffect() is used to log successfully submitted values if there are no form errors:

```
useEffect(() => {

    if (Object.keys(formErrors).length === 0 && isSubmit) {
```

```
      console.log(formValues);
   }
}, [formErrors]);
```

Next, the rules for form data validation are set with the `validateForm` function. This is a simple validation rule that checks whether the `name`, `email`, and `password` form inputs are filled in. It also checks whether the right format is being used for email using `regex.test()`.

For the password, we must check whether the password is more than 8 characters but does not exceed more than 12 characters:

```
const validateForm = (values) => {
    const errors = {};
    const regex = /^[^\s@]+@[^\s@]+\.[^\s@]{2,}$/i;
    if (!values.name) {
      errors.name = "Name is required!";
    }
       ...(This ... represents omitted code which can be
           found on GitHub)
    return errors;
};
```

Then, `onSubmitHandler()` is invoked, which ensures the `setFormErrors()` function is run, which takes `validateForm()` as a parameter. If there are no errors in the form, `setIsSubmit` is set to `true`, allowing the form to be submitted:

```
const onSubmitHandler = € => {
  e.preventDefault();
  setFormErrors(validateForm(formValues));
  setIsSubmit(true);
};
```

Here's the JSX that's returned from the form component with `onSubmitHandler` and each of the errors displayed by the `formErrors` object:

```
return (
  <div className="signUpContainer">
    <form onSubmit={onSubmitHandler}>
      <h2>Create an account</h2>
      <div className="signUpForm">
        <label htmlFor="name">Name</label>
        <p style={{color:'red', fontWeight:'bold'}}>
          {formErrors.name}</p>
        <input
          id="name"
```

```
            type="text"
            name="name"
            value={formValues.name}
            onChange={onChangeHandler}
         />
    ...(This ... represents omitted code which can be found
       on GitHub)
          <button>Register</button>
        </div>
      </form>
    </div>
  );
};
export default SignUp;
```

The following figure shows the output for form entries when form fields are filled before being submitted:

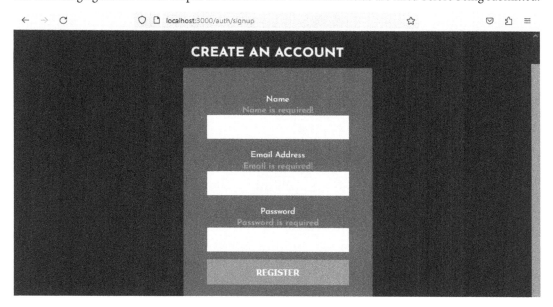

Figure 6.6 – Screenshot of a form with validation rules

We added the `formErrors` property to every form element to output an error message if an error exists. For instance, `{formErrors.name}` displays an error for `name` if a name is not filled in the form.

The full code for this validation rule can be found in this book's GitHub repository: `https://github.com/PacktPublishing/Full-Stack-Flask-Web-Development-with-React/blob/main/Chapter-06/`.

This wraps up how you can add validation to your React application without the use of an external library, thus minimizing the number of potential dependencies your application might rely on.

Summary

In this chapter, we discussed two vital concepts in React: forms and routing. We highlighted the subtle difference in the forms that are set up in non-React and React applications. React provides tons of improvement in how it handles form elements through the use of controlled and uncontrolled form components to enhance the user experience.

Then, we delved into validation concepts and how you can implement validation rules in React. We then spoke about React Router. We demonstrated how React Router, a third-party library, enables us to navigate complex React applications. We discussed the use of `Route`, `Links`, and nested `Routes` and explored how they are used in a React project.

In the next chapter, we will learn about and understand how to implement testing in React applications. Testing is an essential part of software development as it ensures the components of an application work as they should and that the relevant best practices are observed in development.

7
React Unit Testing

We sufficiently discussed the fundamentals of React in the preceding chapters. You have been exposed to React tools and resources to master modern frontend techniques in your full stack development journey. We explored in-depth useful information required to build rich interactive interfaces with React. We discussed the topics of components, props and state, JSX, event handling, forms, and routing in React.

In this chapter, we will focus on unit testing in React applications, a type of software testing that focuses on isolated piece(s) of code. Also, we will explore **Jest**, a Node-based test runner. The test runner allows you to discover test files, run tests, and find out whether tests passed or failed automatically. You will then end with a report in a very clear, expressive, and human-readable format.

Jest is a popular testing framework developed with React in mind. The project was formally owned by *Meta*, the same company behind React. However, with the recent transfer of Jest from Meta to the *OpenJs Foundation*, Jest now has an independent core team working actively on it to ensure stability, security, and code bases that can stand the test of time.

In addition, we will briefly examine a useful testing tool in the React ecosystem – the **React Testing Library (RTL)**. This library provides us with arrays of methods to virtually run tests on React components.

Finally, in this chapter, we will deep-dive into **test-driven development** (TDD), a software development paradigm that places priority on test implementation before actual coding. The coding and testing are interlaced. Testing always comes first, although it is obviously not a fun part of the software development process.

Most experienced developers still struggle with testing just as beginners do. Companies have different policies around testing, but in real-world development, you can't have industrial-strength web applications without testing.

Testing ensures that you follow the best practice of the software development life cycle and that you and your application end users have confidence in your web application functionalities. With appropriate testing, the features of your web application will perform efficiently as expected.

In this chapter, we'll cover the following topics:

- What is software testing?
- Introducing Jest
- Writing tests in Jest
- Unit-testing React components
- TDD

Technical requirements

The complete code for this chapter is available on GitHub at: `https://github.com/PacktPublishing/Full-Stack-Flask-and-React/tree/main/Chapter07`

What is software testing?

Software testing is a process that ensures all the pieces of code in software or application development work as expected by all stakeholders. The process of testing is one of the key phases in software development life cycles that explains the standardized methodology of designing and developing an application.

Software testing ensures that web applications have fewer bugs, technical requirements are implemented as expected efficiently, development costs are reduced, and ultimately, business needs are met.

Interestingly, testing, if carried out effectively, usually gives software engineers the confidence to know full well that the product they ship to production is maintainable, readable, reliable, and well structured.

This creates less panic about possible costly bugs in an application, which may cost a company embarrassment or the erosion of customers' confidence. There are various types of software testing worthy of note – unit testing, integration testing, performance testing, usability testing, acceptance testing, and regression testing.

Let's briefly discuss the types of testing that we have in software development to refresh our minds and have a solid base to build on in the subsequent sections.

- **Unit testing**: In this type of testing, the smallest piece of code or unit in software or application development is tested. In unit testing, you systematically go through three stages – planning, case scripting, and testing. You are primarily concerned with the functionality of the independent units of your application. In the *Unit-testing React components* section, we will deep-dive into unit testing and how you can implement unit tests in your React applications.

- **Integration testing**: In this type of testing, groups of individual units of codes are combined and tested for effective interaction with one another.

- **Performance testing**: In this type of testing, the speed and effectiveness of your application are tested against a given workload. This approach is used to identify early bottlenecks in the smooth running of a software application.

- **Usability testing**: In this type of testing, intended users of your application are allowed to get a feel for your product. The feedback from direct evaluations of your application design through usability testing ensures you are able to capture users' frustration while interacting with your application so early. The report on usability testing can also provide other useful insights that might improve your product.

- **Acceptance testing**: In this type of testing, client satisfaction is gauged – customer requirements are evaluated against a developed software or application. The customer is able to check that those requirements are captured properly and work as expected.

- **Regression testing**: In this type of testing, we check whether code changes occasioned by the addition of new functionality into your software or application do not break down the previously tested working functionality. Regression prevents the introduction of errors or unintended side effects as a result of modifying the code base of your application. With regression testing, all test cases are rerun so that no new bugs are introduced.

Next, we will explore one of the leading testing frameworks you can confidently work with when testing the functionality of your frontend applications.

Introducing Jest

Jest is an open source, robust, well-documented, fast, and safe JavaScript testing framework. Jest provides the tools and resources that you need to have absolute confidence in your JavaScript code base. Testing is fundamental to quality code writing, and Jest with almost zero configuration delightfully handles test implementation.

Jest is very popular in the JavaScript community. You can write tests in Jest leveraging its feature-rich set of APIs, such as matchers, mocked functions, code coverage, and snapshot testing. Jest can be used to test React components, and the React team actually recommends Jest to test React projects.

Jest is a Node.js test runner, which means that the tests always run in a Node environment. Jest is built with performance in mind. You can run a suite of tests in parallel effectively. The Jest toolkit comes with code coverage that enables you to have information about tested and untested files in your project. And when your tests fail, you are provided with insightful information about why they failed.

Let's set up a Jest environment for testing.

Setting up the Jest environment

Node.js is required for Jest to run. Jest comes with the `npx create-react-app` command we used to the `create-react-app` boilerplate code in *Chapter 2, Getting Started with React*.

In this section, we will create a new working folder named `bizza-test` to better explore how Jest works with JavaScript functions, before we move on to using it in the unit testing of React components:

1. Inside `path/to/bizza-test/`, let's execute the following commands in the terminal:

    I.   Run `npm init -y` in the command terminal.

    II. Install Jest as a dependency in the working folder (`bizza-test`) with `npm install --save-dev jest`.

2. Update the `package.json` to be able to run Jest via `npm test`:

    ```
    "scripts": {
                "test": "jest"
    }
    ```

    The complete `package.json` should look like this:

    ```
    {
      "name": "bizza-test",
      "version": "1.0.0",
      "description": "",
      "main": "index.js",
      "scripts": {
        "test": "jest"
      },
      "keywords": [],
      "author": "",
      "license": "ISC",
      "devDependencies": {
        "jest": "^29.2.1"
      }
    }
    ```

The following is a screenshot showing the installation of Jest in the terminal.

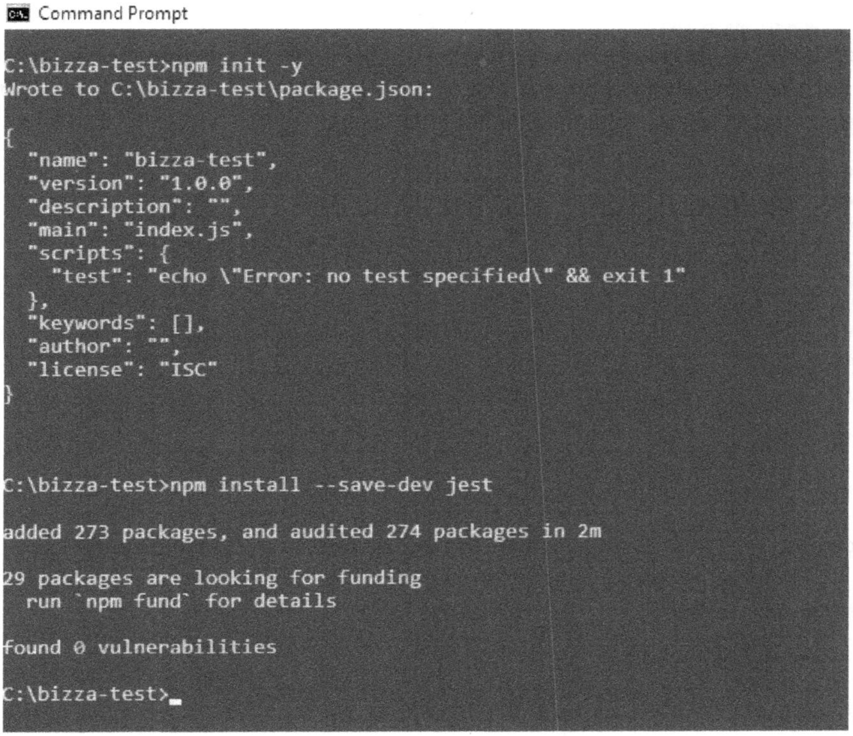

Figure 7.1 – A screenshot of the Jest library setup

With this set, you are on the right path to run your test. Next, we will explore some key testing terminologies and write the actual tests.

Writing tests in Jest

Now that we have set up the test running environment, let's briefly understand some of the keywords we will encounter in this section. In Jest, you have some test keywords and functions from the Jest API to structure your tests; let's examine them:

- `test()` or `it()`: This is a descriptive unit test or test case. It contains three parameters – a descriptive test case string, a function that contains the expectations to test, and an optional timeout argument, specifying how long to wait before aborting a test case. The default timeout is five seconds:

```
test(name, fn, timeout)
test("<test case name>", ()=>{
```

```
                 . . .
    })
```

- `describe()`: This groups related test cases together. It is used to group several related tests and describe their behaviors. `describe()` takes two parameters – a string describing your test group and a callback function that holds test cases:

```
describe(name, fn)
describe("<your test group name>", ()=>{

    test("<test case name>",()=>{
      . . .
    });
    . . .
    test("<test case name>",()=>{
      . . .
    });
})
```

- `beforeAll()`: This function is run before any of the tests in the test file run. And when you have a promise or generator returned from a function, Jest waits for that promise to resolve before running tests:

```
beforeAll(fn, timeout)
```

- `beforeEach()`: This function runs before each of the tests in the test file runs. When you have a promise or generator returned from a function, Jest waits for that promise to resolve before running tests as well.

- `expect()`: In writing a test, you will need to check that a value meets certain conditions at the very least. `expect()` allows you to conditionally check a value with the help of matchers, which we will talk about in a little while. It is noteworthy that there are situations whereby a single test case could have multiple `expect()` functions:

```
test("<test case name>",()=>{
    . . .
    expect(value);
    . . .
})
```

- `Matchers`: Jest uses matchers to test value validation. A matcher lets you test values in different ways to make the right assertion.

- `Assertion`: An assertion is defined as an expression that contains testable logic to verify assumptions made by programmers about a specific piece of code. This allows you to identify errors and other defects in your application.

So, with all these functions and terms well defined, let's write our first test suite:

1. Inside the `bizza-test` working directory, create a `__tests__` folder. Jest searches your working directory for `__tests__` or a test suite file ending with `.test.js` or `.spec.js`, and then it runs the file or all the files in the `__tests__` folder. The `__tests__`, `.test.js`, and `.spec.js` trio are test naming conventions.

   To name the test directory in this book, we will adopt the naming convention `__tests__` case; create a `__tests__` directory inside `path/to/bizza-test/`. This is the folder we will keep the testing files in.

2. Now, create `basicFunctions.js` inside the `bizza-test` directory and add the following snippet to `basicFunctions.js`:

```
const basicFunctions = {
    multiply: (number1, number2) => number1 * number2}
Module.exports = basicFunctions;
```

   This snippet depicts a small JS function that multiplies two numbers together.

3. Inside `__tests__`, create a test suite file, `basicFunctions.test.js`, and paste the following code snippet:

```
const basicFunctions = require ('../basicFunctions');
test ('Multiply 9 * 5 to equal 45', () => {
    expect (basicFunctions.multiply(9, 5)).toBe(45);
});
```

Let's briefly explain the working of the preceding test file code:

- The first line imports the `basicFunctions.js` module we want to test into scope.

- `Test ()` sets the test case description with its function to verify that the multiplication of 9 and 5 equals 45.

- `toBe(45)` lets us know we expect to have 45 as the expected result. If it's anything else, the test will fail.

- Then, inside the Terminal, run the `npm test` command.

The following screenshot shows the output after running the npm test command:

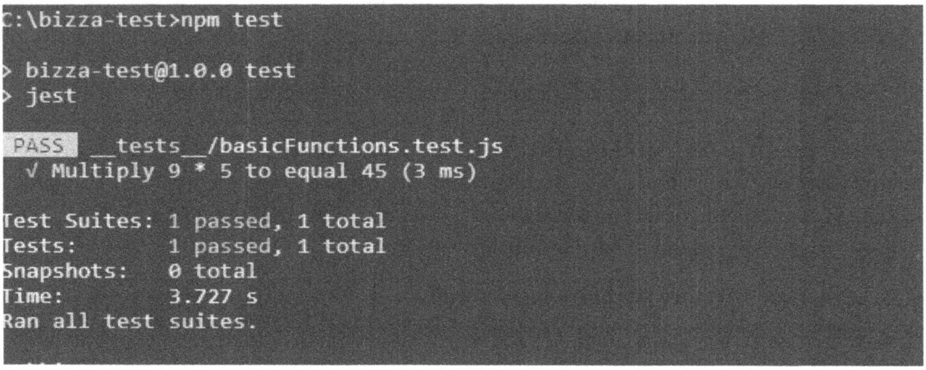

Figure 7.2 – Showing the screenshot for the test case output

The preceding demonstrates how testing works in Jest.

Now, with these solid fundamentals on how we test simple JS functions, let's dive into unit-testing React components.

> **Note**
>
> A list of Jest's useful functions can be found here: https://jestjs.io/docs/api.
>
> A list of commonly used matchers can also be found at https://jestjs.io/docs/expect.

Unit-testing React components

React is hinged on the component-driven development philosophy, and testing a unit of a React component further takes us closer to the basic elements that make up a React component. The essence of unit testing is to test an individual chunk of code to ascertain that the functionality works as expected by the users.

As stated earlier, in unit testing, you systematically go through three stages – planning, case scripting, and testing. Writing a unit test should be thoroughly planned, descriptive test cases should be implemented, and assertions should be clear enough to be understood by everyone on the team.

However, before we delve into unit-testing a React component, how do we know what to test in a React application? It is simple. Every React application has one or more components with specific functions. So, what to test in a React application is subjective. Every project is different and so is the functionality of the applications. The application features to test in an e-commerce web application will be different from features of interest to test in a social media application.

However, there are general rules of thumb for selecting features to test. In application development, testing provides us with confidence and assurance that our software product still works as expected, even with code refactoring. It then boils down basically to the business values of those functions, in addition to those functionalities that significantly impact the end user experience with your application.

To test React components, there is another simple, useful testing tool called RTL. RTL can be used along with Jest to achieve React components testing objectives. RTL lets you test your component units just as a real user of your application would interact with your application UI.

Whereas Jest is a test runner to find tests in a testing environment, run the tests, and affirm whether the tests fail or pass, RTL provides utility functions on top of VDOM. You will come across methods such as `render()` to simulate the rendering of components, `fireEvent` to dispatch a desired event as if a user interacts with the browser, and a screen to query the rendered elements.

RTL also uses the `waitFor()` method to wait for asynchronous codes and `expect()`, which is a function used for making assertions, determining whether the expected outcomes match the actual outcomes, and indicating success or failure. With a React project set up using `npx create-react-app`, you don't have to explicitly install RTL. In a real-world application development environment, you would want to test that your components work as expected. RTL facilitates a simulation of how users would interact with your component.

This is achievable with the in-built utility in RTL that allows you to write tests that interact with DOM nodes directly, without actually rendering React components. Shortly, we will dive into the implementation of how RTL closely mimics how humans can interact with a React application UI in both stateless and stateful components.

Let's start by writing a simple unit test to check whether a component renders.

Writing a unit test for a stateless component

For testing purposes, let's create a fresh React test project with `npx create-react-app bizzatest`. Delete some of the boilerplate files and let the directory structure appear as follows:

```
/bizzatest
    /.git
    /node_modules
    /public
    /src
    /App.js
    /index.js
    /setupTests.js
      .gitignore
        package.json
    package-lock.json
    README.md
```

The App.js file should contain the following code:

```
function App() {
    return (
        <div>
            <h1>Bizza Tests</h1>
        </div>
    );
}export default App;
```

The Index.js file should contain the following code:

```
import React from 'react';
import ReactDOM from 'react-dom/client';
import App from './App';
const root =
    ReactDOM.createRoot(document.getElementById('root'));
root.render(
    <React.StrictMode>
        <App />
    </React.StrictMode>
);
```

Now that we have the testing environment set up, let's create a stateless component and write a unit test to check whether WelcomeScreen.js renders the **Welcome to Bizza Conference Platform** paragraph text. Even without actually mounting the component, we can test whether WelcomeScreen. js will render the intended text paragraph.

Follow the proceeding steps to create the component, and unit-test it for the presence of the specified paragraph text.

Creating a WelcomeScreen.jsx component

Let's create a src/WelcomeScreen.jsx component and add the proceeding code snippet. This component displays the <h1>React</h1> element:

```
import React from "react";
const WelcomeScreen = () => {
    return   <h1>React</h1>
};
export default WelcomeScreen;
```

Creating a test case file

Inside the `src/__tests__` folder, create a test case file and name it `WelcomeScreen.test.js`. Use it to store all your test cases. You can also write your test in each of your component's folders. However, in this case, we will store it inside the `src/__tests__` folder. Add the following code snippet:

```
import {render,screen,cleanup} from
    "@testing-library/react";
import WelcomeScreen from "../WelcomeScreen";
afterEach(() => {
    return cleanup();
});
test("should show Welcome text to screen", () => {
    render(<WelcomeScreen />);
    const showText = screen.getByText(/Welcome to Bizza
        Conference Platform/i);
    expect(showText).toBeInTheDocument();
});
```

Let's briefly discuss what is happening in the preceding snippet.

- The `render`, `screen`, and `cleanup` utilities are imported from `@testing-library/react`. `render` helps you to virtually render a component in a container that is appended to the body of the HTML document (DOM).

  The `screen` object from RTL provides you with methods to find the rendered DOM elements to make the necessary assertions. Essentially, `screen` is used to interact with rendered components, and the `cleanup` function is used to clean up the rendered component after each test.

- The `WelcomeScreen` component is imported as a required file for the test. This is the component to be tested.

- The `afterEach()` method is added to unmount every rendered component test before the next component is rendered to prevent memory leaks. The `afterEach()` block is a Jest life cycle method that runs after each test. In this case, it calls the `cleanup` function to clean up any rendered components after each test. Using this cleanup utility from TRL is regarded as a best practice in React component testing.

- The `test()` function defines the actual test, named `"should show Welcome text to screen"`, and a callback function to hold the test case. The test first calls the `render` function to render the `WelcomeScreen` component. Then, it uses the `screen.getByText` function to get the DOM element that contains the text **Welcome to Bizza Conference Platform**. The `expect()` function is then used to verify that the text is in the document, using the `toBeInTheDocument` matcher.

- When the `WelcomeScreen` component is rendered, we expect it to contain **Welcome to Bizza Conference Platform**.

Now, run the test with `npm test`.

The following screenshot shows the output:

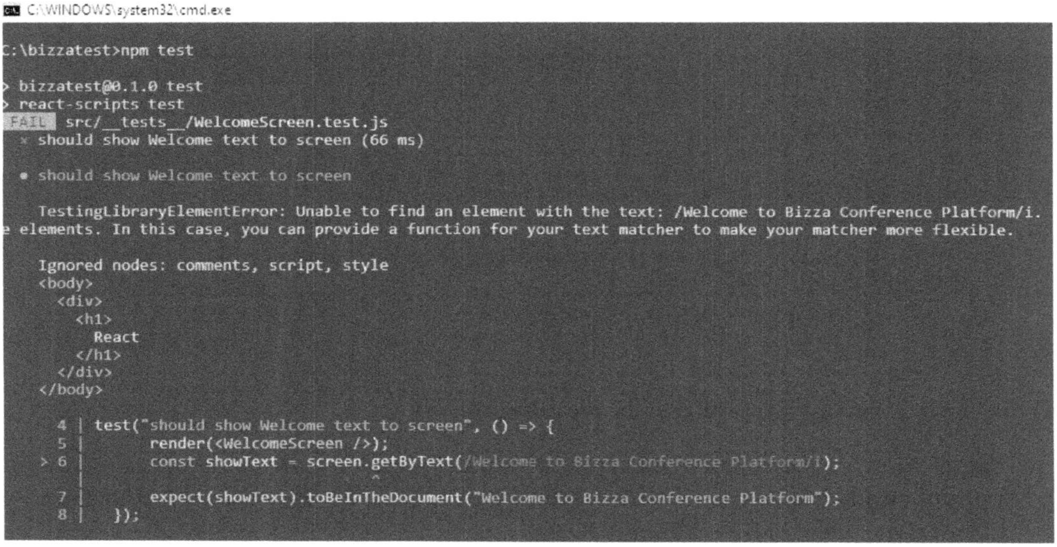

Figure 7.3 – A screenshot showing failed test case output

As expected, the test failed. The paragraph text in the RTL-simulated rendered container in `WelcomeScreen.jsx` is `<h1>React</h1>` and not `<h1>Welcome to Bizza Conference Platform</h1>`.

Updating the WelcomeScreen component

Let's now update `WelcomeScreen.jsx` with the expected text to be rendered on the screen:

```
import React from "react";
const WelcomeScreen = () => {
    return <h1>Welcome to Bizza Conference Platform</h1>
};
export default WelcomeScreen;
```

Awesome, the test passed! The updated test report now shows a passed test:

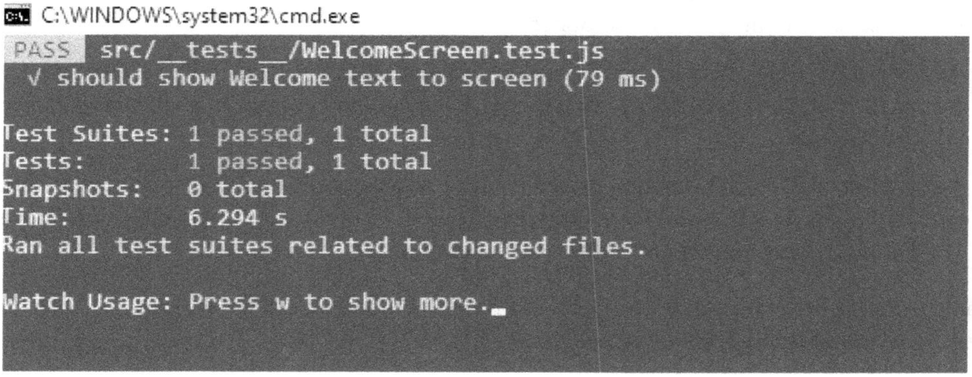

Figure 7.4 – A screenshot showing a passed test case output

You can write unit tests for the elements of your various components and use the appropriate queries from RTL. These RTL queries allow you to find elements in the same way that users of your application will interact with your UI:

- `getByRole()`: This function is used to locate an element by its role attribute, such as button, link, checkbox, radio, and heading. The `getByRole()` function is useful for testing the accessibility of a component, as well as for general testing purposes.

- `getByLabelText()`: This function is used to locate a form element that is associated with a label element using the `for` attribute. `getByLabelText()` is useful for testing form components and their accessibility.

- `getByPlaceholderText()`: This function is used to locate an input element by its placeholder attribute. `getByPlaceholderText()` is useful for testing input fields and their behavior.

- `getByText()`: This function is used to locate an element by its text content. `getByText()` is useful for testing the rendering of specific text or locating buttons or links that are identified by their text.

- `getByDisplayValue()`: This function is used to locate a form element by its displayed value, such as an input or select element that has a pre-filled value. `getByDisplayValue()` is useful for testing form components and their behavior.

- getByAltText(): This function is used to locate an image element by its alt attribute, which provides a textual description of the image. getByAltText() is useful for testing the accessibility of images in a component.

- getByTitle(): This function is used to locate an element by its title attribute. The getByTitle() is useful for testing the accessibility and behavior of elements that have a title attribute.

- getByTestId(): This function is used to locate an element by its data-testid attribute. getByTestId() is useful for testing specific elements that are identified by a unique test ID, without relying on other attributes such as class or ID.

Let's now examine how you can write a unit test for a stateful component. We will make use of a speaker card and test its units to our satisfaction.

Writing a unit test for a stateful component

In the components folder, create a stateful component named SpeakerCard. The SpeakerCard component is a functional component that renders a card with information about a speaker. The component takes a speaker object as a prop, which contains properties such as the speaker's name, occupation, company, phone number, and email address.

Now, the following snippet shows the test snippet for the SpeakerCard component:

```
import {useState} from 'react'
const SpeakerCard=speaker=>{
const {name, occupation, company, phone, email}= speaker;
const [showDetails, setShowDetails]= useState(false);
const toggleDetails = () => setShowDetails(!showDetails);

return(
<div className="card" data-testid="card">
<span><h2>Name:{name}</h2></span>
<span><h2>Occupation:{occupation}</h2></span>
<span><h2>Company:{company}</h2></span>
<button data-testid="toggle-test" onClick={toggleDetails}>
{showDetails? "Hide Details":"Show Details"}
</button>
{showDetails && (<div data-testid="test-details">

<h2>Email:{email}</h2>
<h2>Phone:{phone}</h2>
</div>
)}
</div>
```

```
    )
  }
export default SpeakerCard;
```

The preceding code snippet presented is explained here:

- The component takes a `speaker` object as a prop, which contains properties such as the speaker's name, `occupation`, `company`, `phone number`, and `email address`.

- The component uses the `useState` hook to manage a Boolean state variable called `showDetails`, which controls whether additional details about the speaker are shown or hidden. The `toggleDetails` function toggles the value of the `showDetails` variable when the toggle button is clicked.

  If the `showDetails` variable is `true`, the component renders additional details about the speaker, including their email and phone number, within a nested `div` element with a `data-testid` attribute of `test-details`.

- The `toggleDetails()` function's initial state is `false`; when the toggle button is clicked, the state changes to `true`. The `data-testid` attribute in `<div data-testid="card">` is used to identify the DOM node that you test. The `data-testid` attribute has `card` as its value. This value allows `expect()`, a Jest utility, to make assertions about whether the test fails or passes.

- With the toggle button, we set the `data-testid` attribute to `toggle-test` to assert that no button has been clicked yet.

- The `data-testid="test-details"` is used to assert that the toggle button is clicked and details are displayed on the screen.

- Conditionally, when `showDetails` is set to `true`, email and phone details are displayed on the screen; otherwise, they would be hidden.

Now, let's write the unit test to show that `<SpeakerCard/>` can render on screen and when the toggle button is clicked, we can see more details about the `speaker` object.

Inside `/src/__tests__/`, create a test file, `SpeakerCard.test.js`:

```
import {render, screen, fireEvent, cleanup} from
    "@testing-library/react";
import SpeakerCard from
    '../components/SpeakerCard/SpeakerCard'

const speaker= {
    name: "Olatunde Adedeji",
    occupation: "Software Engineer",
    company: "Mowebsite",
```

```
        email:"admin@admin.com",
        phone: "01-333333",
    }
afterEach(() => {
    return cleanup();
});

test("should render the SpeakerCard component", ()=>{
    render(<SpeakerCard/>);
    const card = screen.getByTestId("card");
    expect(card).toBeDefined()
});

test("should make sure the toggle button shows or hides
    details", ()=>{
    render(<SpeakerCard speaker={speaker}/>);
    const toggleButton = screen.getByTestId("toggle-test");
    expect(screen.queryByTestId(
        "test-details")).toBeNull();
    fireEvent.click(toggleButton);
    expect(screen.getByTestId(
        "test-details")).toBeInTheDocument();
});
```

Let's go over the preceding code snippet:

- **Test 1**: This should render the SpeakerCard component.

    I.   Import SpeakerCard.jsx as a required file for the test.

    II.   Define the test() function from the @testing-library/react library with
          a test string describing what the test should do, which in this test case is "should
          render the SpeakerCard component", and a function that contains the test code.

    III. Then, import and use the render and screen utilities to simulate the rendering of
          the <SpeakerCard/> component.

    IV.   The SpeakerCard component accepts the props defined speaker object.

    V.   Query the node with .getByTestId("card") and assign its value to card. This
          allows you to access the DOM node needed to make an assertion. You can then use
          expect() from the Jest utility to ascertain that <SpeakerCard/> is rendered.
          Expect it to be defined!

The following screenshot shows a passed test for the rendering React component:

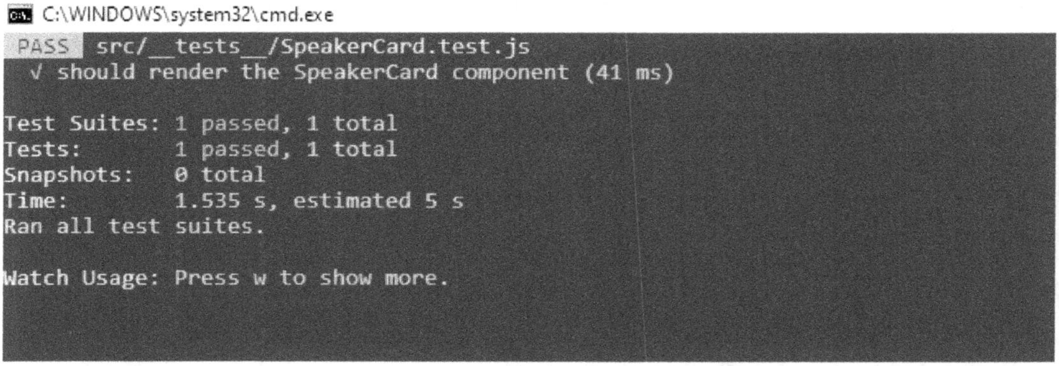

Figure 7.5 – A screenshot showing a passed rendering component test

- **Test 2**: This should make sure the toggle button shows or hides details:

  I.   Define the `test()` function from the `@testing-library/react` library with a `test` string describing what the test should do, which in this test case is `"should make sure the toggle button shows or hides details"`, and a function that contains the test code.

  II. Render the `SpeakerCard` component.

  III. When no button is clicked, we expect the `data-testid` attribute value `toggle-test` to be *n*. The first assertion checks that the details section is not initially displayed, by checking that the element with `test-details data-testid` is not in the document. This is done using the `screen.queryByTestId()` function, which returns `null` if the element is not found.

  IV. Then, import and use the `fireEvent` function. The `fireEvent.click(toggleButton)` simulates a user clicking on a toggle button in the `SpeakerCard` component. The `fireEvent` utility is a part of the `@testing-library/react` package, which provides a way of simulating user interactions in a test environment. The `fireEvent.click()` function is used to simulate a click on the toggle button. This will trigger the `toggleDetails` function in the component, which should show or hide the details section based on the `showDetails` state.

  V.   Query the node with `getByTestId` and assign its value to `toggleButton`. The `data-testid` attribute in the `SpeakerCard` component checks whether the details are displayed or not by searching for the `test-details` element using `screen.getByTestId`.

VI.   We expect `test-details` from `data-testid` to display on the screen. If `test-details` is present in the document, the test passes; otherwise, the test fails.

The following screenshot shows the testing of the toggle button:

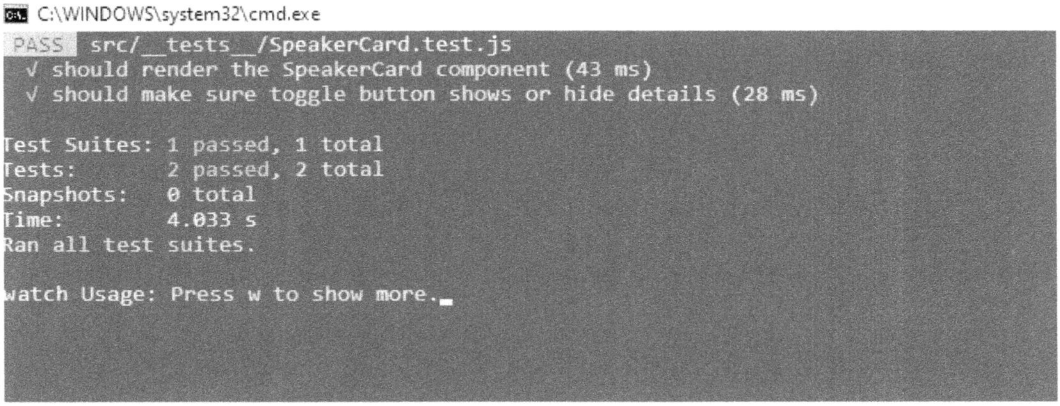

Figure 7.6 – Screenshot showing the passed toggle button test

Next, we will discuss TDD. This is a software development practice that encourages the testing of every functional unit of an application first before coding it.

TDD

TDD is a development paradigm that puts writing tests first. You write the test first and then write code to validate. The main purpose of TDD is rapid feedback. You write a test, run it, and it fails. You then write minimal code to pass the test. Once the test passes, you then refactor your code appropriately.

These processes are iteratively repeated. Focusing on writing tests before code implementation allows developers to see the product from the users' point of view, thus ensuring a working functionality that meets the users' needs.

TDD enables software developers to come up with units of code base with a single responsibility – allowing code to do just one thing that works properly. However, the traditional approach is to code and then test. The idea of testing a code base at the end of the development process has been proven to be flawed and comes with a high cost of code maintenance.

Most software developers are more agile than test-driven. The urgent need to get a product to market with often unrealistic deadlines places less priority on the quality of the units of code powering such software products. With the traditional approach to development, you have errors sneaking into production.

As we all know, web applications in production do have additional features added from time to time based on business demands. However, without quality tests, you can have a situation where new feature addition or fixes create more problems.

Another problem with the traditional approach is having a software development team for a product different from the testing team. This separation of teams can make the long-time maintenance of the code base difficult and may result in a terrible decline in code quality, owing to a chaotic clash of intents.

With TDD, you and other stakeholders have proof that your code does work and is of high quality. Right now, we will wear the hat and shoes of a web developer who is agile and simultaneously test-driven. We will examine a case study where we build a login component as part of our *Bizza* project using a TDD methodology.

The component development will comply with a TDD principle:

1. Write a single test case. This test will be expected to fail.
2. Write minimal code that satisfies the test and makes it pass.
3. Then, refactor your code and run it to pass/fail.

These steps are repeated.

So, now let's write our initial tests. We expect them to fail anyway. This is the essence of TDD.

Writing a single test case

In the `bizzatest` directory for testing, create the `bizzatest/src/components/Auth/SignInForm/SignInForm.jsx` component and update the file as follows. This is a `SignInForm` component with no labels and functionalities:

```
import React from 'react';
const SignInForm = () => {
    return (
        <div className="signInContainer">
            <form>
                <div className="signInForm">
                    <label htmlFor="email"></label>
                    <input
                        type="email"
                    />
                    <label htmlFor="password"></label>
                    <input
                        type="password"
                    />
                    <button></button>
                </div>
```

```
                </form>
            </div>
        );
    };
    export default SignInForm;
```

Inside App.js, add the following code snippet:

```
import React from "react";
import "./App.css";
import SignInForm from
    "./pages/Auth/SignInForm/SignInForm";
const App = () => {
    return <SignInForm />;
};
export default App;
```

Then, run npm s to render the following component:

Figure 7.7 – A screenshot showing a rendered SignInForm component

In TDD, you want to start with a unit test of a component that will fail, and then work around to make it pass after the actual development of the component features.

> **Note**
> Check the *Technical requirements* section for this chapter to get the styles for this form component (https://github.com/PacktPublishing/Full-Stack-Flask-and-React/tree/main/Chapter07/05).

Now, let's write and run the test for the SignInForm component.

Create the `test` file inside `bizzatest/src/__tests__/ SignInForm.test.js`.

Update the file as follows:

```
import { render, screen,cleanup } from
    '@testing-library/react';
import  SignInForm from
    '../components/Auth/SignInForm/SignInForm';

afterEach(() => {
    return cleanup();
  });

test("Email Address should be rendered to screen", () => {
    render(<SignInForm />);
    const linkElEmailInput =
        screen.getByText(/Email Address/i);
    expect(linkElEmailInput).toBeInTheDocument();

});

test("Password should be rendered to screen", () => {
    render(<SignInForm />);
    const linkElPasswordInput =
        screen.getByText(/Password/i);
    expect(linkElPasswordInput).toBeInTheDocument();   });

test("Sign In should be rendered to screen", () => {
    render(<SignInForm />);
    const linkElSignInBtn = screen.getByTest(/SignIn/i);
    expect(linkElSignInBtn).toBeInTheDocument();
});
```

Let's explain the workings of the preceding snippet:

`Import { render, screen } from '@testing-library/react'` brings to scope the `render` and `screen` functions in RTL. This allows us to render components with VDOM in the testing environment.

The required `SignInForm` component file under test is then imported.

`test("Email Address should be rendered to screen", () => {render(<SignInForm />)` defines a test case with a description and renders `SignInForm`-component. The `render()` method comes from the RTL. The same thing is repeated for the `Password` and `Sign In` buttons with the individual test case descriptions.

In the first test, the test checks that the `Email Address` label is rendered on the screen by using the `getByText()` method from `@testing-library/react`. The test passes if the label is found on the screen.

In the second test, the test checks that the `Password` label is rendered on the screen by using the `getByText()` method. The test passes if the label is found on the screen.

The third test checks that the button with the text `SignIn` is rendered on the screen by using the `getByTestId()` method. The test passes if the button is found on the screen.

`expect(linkElEmailInput).toBeInTheDocument()` is for the assertion. This is to verify that the values of the declared variables are present in the `SignInForm` component.

Now, in the Terminal, run the `npm test` command. The test case descriptions are shown here with failed statuses.

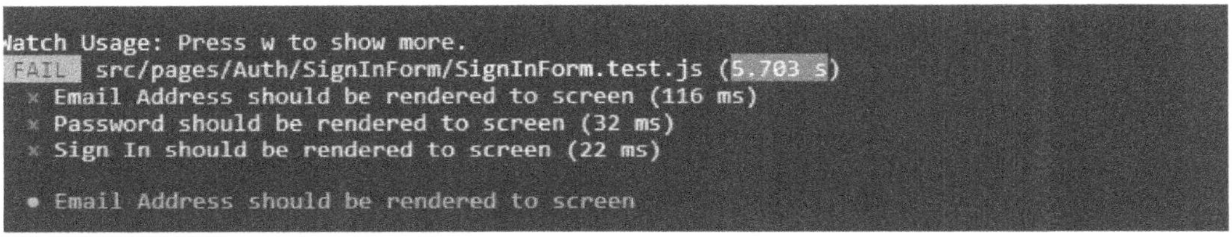

```
Watch Usage: Press w to show more.
 FAIL  src/pages/Auth/SignInForm/SignInForm.test.js (5.703 s)
  × Email Address should be rendered to screen (116 ms)
  × Password should be rendered to screen (32 ms)
  × Sign In should be rendered to screen (22 ms)

  ● Email Address should be rendered to screen
```

Figure 7.8 – A screenshot of a failed SignInForm test

The following screenshot shows a detailed report of the Jest test runner. It shows one test suite and three failed tests.

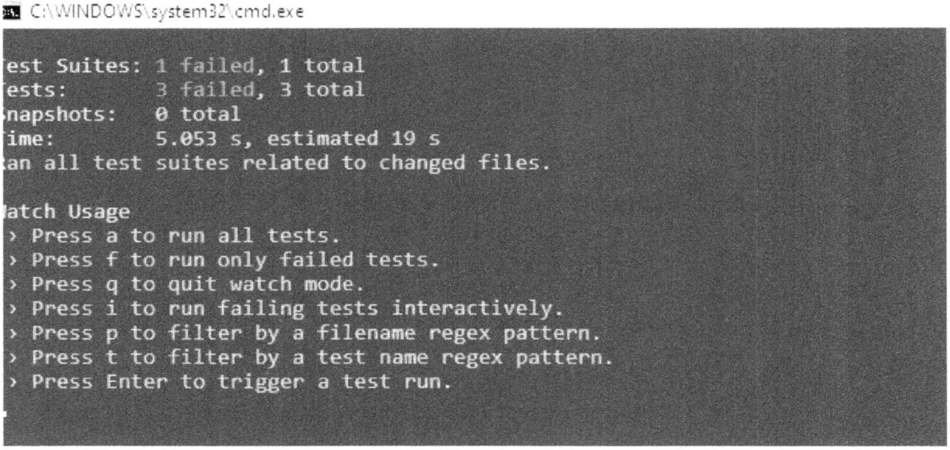

C:\WINDOWS\system32\cmd.exe

```
Test Suites: 1 failed, 1 total
Tests:       3 failed, 3 total
Snapshots:   0 total
Time:        5.053 s, estimated 19 s
Ran all test suites related to changed files.

Watch Usage
 › Press a to run all tests.
 › Press f to run only failed tests.
 › Press q to quit watch mode.
 › Press i to run failing tests interactively.
 › Press p to filter by a filename regex pattern.
 › Press t to filter by a test name regex pattern.
 › Press Enter to trigger a test run.
```

Figure 7.9 – A screenshot of the failed SignInForm component test

Writing minimal code that satisfies the test and making it pass

Now, update the `SignInForm` component to meet the expectations of the test cases in the test suite:

```jsx
import React from 'react';
const SignInForm = () => {
    return (
        <>
            <div className="signInContainer">
                <form>
                    <div className="signInForm">
                        <label htmlFor="email">
                            Email Address</label>
                        <input
                            type="email"
                        />
                        <label htmlFor="password">
                            Password</label>
                        <input
                            type="password"
                        />
                        <button>Sign In</button>
                    </div>
                </form>
            </div>
        </>
    );
};
export default SignInForm;
```

In the preceding snippet, we refactored the code to pass the test as expected and complete the principle of TDD. The Jest test runner runs automatically and passes the test based on the refactoring of the `SignInForm` component.

In the following figure, we have the detailed Jest success report of our test-driven component development.

```
C:\WINDOWS\system32\cmd.exe

Watch Usage: Press w to show more.
 PASS  src/pages/Auth/SignInForm/SignInForm.test.js (5.411 s)
  √ Email Address should be rendered to screen (111 ms)
  √ Password should be rendered to screen (24 ms)
  √ Sign In should be rendered to screen (222 ms)

Test Suites: 1 passed, 1 total
Tests:       3 passed, 3 total
Snapshots:   0 total
Time:        70.888 s
Ran all test suites related to changed files.

Watch Usage: Press w to show more._
```

Figure 7.10 – A screenshot of a successful pass of the SignInForm component test

The `SignInForm` component now appears, as shown in *Figure 7.11*:

Figure 7.11 – A screenshot of a rendered SignInForm component after passing the test

Code refactoring

Code refactoring is the process of modifying existing code to improve its quality, maintainability, and efficiency without changing its behavior. Code refactoring involves analyzing and improving the code structure, design, and implementation to make it easier to understand, modify, and maintain. Refactoring is typically done to improve code readability, remove code duplication, and increase performance.

During code refactoring, it is important to test your code to ensure that any changes made do not affect the behavior of the system. The code is run multiple times after each refactoring step to ensure that it still passes all the positive tests or fails all the negative tests.

If a test fails when you expect it to pass, that means that the refactoring step introduced a bug and the code needs to be reverted to its previous state, or additional changes need to be made to fix the issue. Refactoring can be done manually, but there are also automated tools such as Jest, RTL, Enzyme, and React testing utilities that can help identify areas of the code that need refactoring.

Refactoring as a part of TDD methodology allows you to refactor your code and run it to pass/fail until you are confident in the test results. These code refactoring steps can be repeated as many times as desired.

Let's refactor `SignForm.test.js` and examine one of the ways you can refactor your test code.

Create the `test` file inside `bizzatest/src/__tests__/ RefactorSignInForm.test.js`:

Update the file as follows:

```
import { render, screen, cleanup } from
    '@testing-library/react';
import SignInForm from
    '../components/Auth/SignInForm/SignInForm';
describe('SignInForm component', () => {
    afterEach(() => {
        cleanup();
    });
    it('should render the email address input', () => {
        render(<SignInForm />);
        const emailInput =
            screen.getByLabelText(/Email Address/i);
        expect(emailInput).toBeInTheDocument();
    });

    it('should render the password input', () => {
        render(<SignInForm />);
        const passwordInput =
            screen.getByLabelText(/Password/i);
        expect(passwordInput).toBeInTheDocument();
    });

    it('should render the sign in button', () => {
        render(<SignInForm />);
        const signInButton =
            screen.getByRole('button', { name: /Sign In/i
            });
        expect(signInButton).toBeInTheDocument();
    });
});
```

Here are some of the changes that were made in the refactored test code:

- The tests were wrapped in a `describe` block to group them under the same heading
- `test()` was replaced with `it()` for consistency
- The text matching was changed from `getByText` to `getByLabelText` for the email and password inputs, as this is a more appropriate way to target form inputs
- The `getByTest` query for the `Signin` button was replaced with `getByRole` and a name option

Summary

Testing is crucial to the successful deployment of an application product into production. In this chapter, we examined various testing types available, especially unit testing for code base quality and ease of maintainability. This ensures a lower cost of producing software with confidence.

We also explored Jest, a Node test runner that delightfully tests JavaScript code and, by extension, React applications. The Jest testing framework ensures that you work in an integrated testing environment, with virtually all your testing tools in one place and only a stone's throw away.

We discussed RTL with its implementation and then, in depth, the unit testing of React components using Jest and RTL, which comes with the **Create React App** (**CRA**) boilerplate code.

We wrote useful component tests to showcase the ability of the combined tools of Jest and TLR. Finally, we discussed a testing methodology, TDD, in software development and how it can be used in React-based applications with examples.

Next, we will temporarily shift the focus of this book and delve into the backend development aspect of full stack web development, which is what this book is all about. We will begin a backend application in the next chapter, discussing SQL and data modeling.

Part 2 – Backend Development with Flask

Welcome to *Part 2* of our book, where we will delve into the dynamics of backend development with Flask. In this section, we will explore the fundamental concepts and techniques used in modern web development using the Flask framework.

You will gain the knowledge and tools necessary to tackle the complexities of backend development. From setting up your Flask environment to designing efficient RESTful APIs, implementing database integration, and integrating React frontends with Flask, we will cover it all.

By the end of this part, you will have a solid grasp of Flask's core concepts and be well on your way to becoming a proficient full stack web developer. Let's embark on this exciting journey together!

This part has the following chapters:

- *Chapter 8, SQL and Data Modeling*
- *Chapter 9, API Development and Documentation*
- *Chapter 10, Integrating React Frontend with the Flask Backend*
- *Chapter 11, Fetching and Displaying Data in a React-Flask Application*
- *Chapter 12, Authentication and Authorization*
- *Chapter 13, Error Handling*
- *Chapter 14, Modular Architecture – The Power of Blueprints*
- *Chapter 15, Flask Unit Testing*
- *Chapter 16, Containerization and Flask Application Deployment*

8

SQL and Data Modeling

So far, we've explored React, a key library in frontend technology stacks. Now, we will explore the world of backend development, starting with **Structured Query Language** (**SQL**) and data modeling.

SQL and data modeling are critical skills for any backend developer, and starting with these skills in the backend development journey will give you a strong foundation to build robust and scalable web applications. SQL is a standard language used to manage data in relational databases, which are essentially used to store structured data.

Knowledge of SQL will help you to write optimized SQL queries that can improve the performance of your application and reduce server load. Data modeling will help you to design a database schema that reflects the data your application will work with. Data modeling can help you avoid performance issues, maintainability problems, and other common issues that may arise when working with databases.

We will dive deeply into relational databases and how database tables can relate to one another. We will examine SQL as a standard language used by many database management systems, including **PostgreSQL**, a popular open source **relational database management system** (**RDBMS**).

Having a robust understanding of relational databases and knowledge of how database relationships work will help you design scalable and maintainable data structures. We will leave no stone unturned, cruising through the setup of PostgreSQL and exploring how Flask applications can communicate with PostgreSQL.

We will also discuss in depth the ability of SQLAlchemy to handle various relational database dialects by providing an interface that allows SQL interaction with a database using Python objects. **SQLAlchemy** is an industrial-strength object-relational mapper that provides a powerful interface for interaction with various relational database dialects, including PostgreSQL.

SQLAlchemy makes it easier to write complex database queries and manage database transactions. We will examine how you can come up with a data model and send data from a Flask application to a database.

In addition to data modeling and SQL, migration is also a crucial aspect of backend development. We will examine migration as a way of tracking and updating database structures with Alembic. **Alembic** is a migration tool that provides a reliable way of tracking and updating database structures, making it an essential tool for Python developers.

In this chapter, we will cover the following topics:

- What is the relational data model?
- Exploring the different database relationships
- Setting up PostgreSQL
- Understanding database concepts for Flask applications
- Understanding SQLAlchemy ORM basics
- Modeling data for the speakers conference web application
- Sending data to the PostgreSQL database from a Flask application
- Migration with Alembic

Technical requirements

The complete code for this chapter is available on GitHub at: `https://github.com/PacktPublishing/Full-Stack-Flask-and-React/tree/main/Chapter08`.

What is the relational data model?

The **relational data model** is a conceptual approach used to represent a database as a group of relations. Most web applications are highly data-driven. Developers have to deal with either code-level data storage, in the case of a data structure, or find a way to persistently store data in an RDBMS, such as PostgreSQL or MySQL.

In an RDBMS, you can refer to a table as a relation. Therefore, a relational model represents data as a collection of relations or tables. Breaking down the database structure further, you then have rows and columns making up a table. Then, you have a record, which consists of a combination of rows and columns.

Let's take a look at a hypothetical table named `customers` that depicts the structure of a typical table for clarity:

Id	firstname	lastname	email	phone
1	Joel	Doe	Joel@admin.com	404-228-5487
2	Edward	Spinster	Edward@admin.com	403-268-6486
3	Mabel	Emmanuel	Mabel@admin.com	402-248-4484

Table 8.1 – A table showing customer information

In the preceding `customers` table, we have five columns and three rows. Each row in a table is called a tuple. The column headers such as `Id`, `firstname`, `lastname`, `email`, and `phone` are called attributes or fields. In a database, tables are created to store data efficiently. Each table usually represents a business entity or object such as a speaker, venue, subject, customer, product, order, and so on.

For clarity, a business entity represents things we intend to encapsulate in the application business data model with all the rules, relationships, and ability to be persistent in the database. In the preceding table, we have a customer business entity with the following attributes – `Id`, `firstname`, `lastname`, `email`, and `phone`. More often than not, you would have more than one table in your web application.

Expectedly, you need to be able to relate these different tables or relations in your database using primary keys. The primary key is the unique identifier for the entire row of a table, referring to one or more columns. And if there are multiple columns for the primary key, then the set of primary key columns is known as a composite key. In our case, `Id` in the customer table could be set as the primary key.

Additionally, another concept in data relation is a foreign key. A foreign key refers to a primary key in another (foreign) table. Foreign keys are used to map relationships between tables or relations. The table relationship helps you to store data efficiently where it needs to be stored, and accessing related data becomes easier when needed.

There are many relationships in database design – *one-to-one*, *one-to-many*, and *many-to-many*. For instance, related tables in a database can help you find out orders made by your customers, how many conference attendees are currently enrolled in each subject, and so on.

In the next section, we will extensively discuss the relationships in a database and how they are used in web application architectural design.

Exploring the different database relationships

In the client-server model, the database resides at the server end of the infrastructure. The database is core to any production-grade web application in collecting and storing application data. Understanding relationships that exist in a database is vital for organizing, managing, and retrieving useful data from a database.

As previously mentioned, there are three types of relationships that exist in a database – *one-to-one*, *one-to-many*, and *many-to-many*. We will begin by delving into the concept of a one-to-one relationship.

One-to-one (1:1) relationship

A **one-to-one relationship** in a data model refers to a direct link relationship that exists in information between two tables. With a one-to-one relationship, you have a situation where a record in one table is directly associated with a specific row in another table. For clarity, let's quickly dive into a scenario where you have two tables – `speakers` and `speakerExtras`.

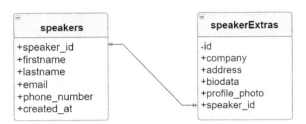

Figure 8.1 – An entity relationship diagram showing the one-to-one relationship

The preceding diagram depicts the speakers table and the speakerExtras table. The speakers table details the basic required information of a speaker object, while the speakerExtras table showcases the optional information that can be added to speaker information. speaker_id in the speakers table is the primary key that can be used to uniquely identify a speaker record.

In the speakerExtras table, a column is added as speaker_id. This speaker_id column in the speakerExtras table is a foreign key with a unique attribute. This speaker_id column is used as a reference key for speaker_id in the speakers table to form a one-to-one relationship between the speakerExtras table and the speakers table.

Most of the time, the one-to-one relationship data model can actually be merged. Merging a one-to-one relationship data model typically means combining the tables into a single table. This is usually done when the two tables have a strong connection and the data in the tables is not complex enough to warrant having separate tables.

However, there are cases where the database design requirement may be a separate table for information that may be optional, so rather than having an empty column in a table, you can create a different table to handle this optional information for better database performance.

Let's take a look at another database relationship, one-to-many.

One-to-many (1:M) relationship

A **one-to-many relationship** in a data model explains a relationship between two tables, in which a row in one table can reference one or many rows in another table. The one-to-many relationship is a kind of parent-child relationship where a child record can only have one parent record.

Most of the time, the parent record has more than one child in another table's rows. However, in a real-life scenario, we could have a situation in which a parent record has no child record at all. What does a parent with no child record mean? It means that the foreign key column in the child table will be empty for that parent record.

For instance, consider a database design for a store where each customer can have multiple orders. If a new customer record is created but no orders have been placed yet, there will be a customer record with no corresponding order records. In this case, the customer record is the parent record, and the order records are the child records.

Furthermore, a one-to-many relationship in some cases also allows a single record in another table. There are cases where a one-to-many relationship can be constrained to functionally act as a one-to-one relationship. For instance, you could add a unique constraint to the foreign key column in the child table, ensuring that each record in the parent table is associated with, at most, one record in the child table.

This would effectively constrain the relationship to function as a one-to-one relationship, even though the underlying data model is a one-to-many relationship. This one-to-many relationship is almost similar to a one-to-one relationship but with a subtle difference of a unique constraint. Why is this one-to-many relationship important in data modeling?

Like other database relationships, the one-to-many relationship is a fundamental building block of relational database design and is essential for organizing, managing, and analyzing complex data structures. The one-to-many relationship enforces referential integrity of data by ensuring that each record in the child table is associated with a valid record in the parent table. This approach helps prevent data inconsistencies and errors that can arise from orphan records or references to non-existent records.

Let's take a look at an example of a one-to-many relationship. Imagine we have `customers` and `orders` tables in our database. It is possible for a customer to have many 1:M orders with a business over a period of time.

A business might want to keep this record. For example, customers tend to have different orders; a particular `orderID` can't belong to many customers. Every customer's order is unique. The following diagram depicts a one-to-many relationship between customers and orders.

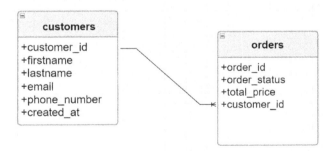

Figure 8.2 – An entity relationship diagram showing the one-to-many relationship

In the preceding diagram, `customer_id` in the `orders` table represents the foreign key and the main linking factor between the two entities – `customers` and `orders`. If you want to know the number of orders a particular customer has, all you need to do is write a `JOIN` query that tends to look up `customers` and `orders` tables, and check for the occurrence of `customer_id` in the `orders` table as a reference key.

In the database design for your web application development, you will encounter more of this data relationship, as it is the most commonly used one. Next, we take a look at another database relationship, many-to-many.

Many-to-many (M:M) relationship

A **many-to-many relationship** in data modeling simply refers to a data relationship in which multiple records in one table are related to multiple records in another table. Take, for instance, a conference database design, where you could have the following scenarios:

- An attendee can be enrolled in multiple conference sessions
- A session can have many attendees enrolled in it

This implies that an attendee has many conference sessions, and a conference session has many attendees. Thus, there is a many-to-many relationship between the conference attendees and sessions.

Interestingly, unlike a one-to-one relationship and one-to-many relationship, you can't model a many-to-many relationship with just two tables. In many-to-many relationship modeling, a third join table is required, whereby you have the values of the primary keys of the two tables added to the join table.

Let's examine the **entity relationship** (ER) diagram of a many-to-many relationship for conference attendees and the sessions they enroll in:

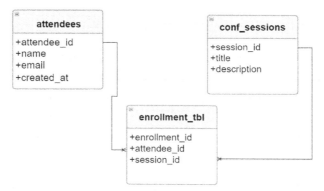

Figure 8.3 – An entity relationship diagram showing the many-to-many relationship

Let's break down the preceding ER diagram to have a better understanding of the many-to-many relationship. We have three tables, as clearly shown – `attendees`, `enrollemnt_tbl`, and `conf_sessions`.

The `attendees` table contains records for conference attendees. Likewise, the `conf_sessions` table contains records for conference sessions. You then have a join table, `enrollment_tbl`, which technically forms a one-to-many relationship with each of the two tables.

enrollment_tbl contains the primary keys of both the attendees and the conf_sessions tables as foreign keys. With enrollment_tbl, we can query related records from the attendees and conf_sessions tables. In this case, we can access all the sessions a particular attendee attends at the conference.

Next, we will delve deeper into database setup, using SQLAlchemy and Alembic, for the database part of a web application.

Setting up PostgreSQL, SQLAlchemy, and Alembic

We are going to start by setting up the database tools needed for the backend database. PostgreSQL is a popular free and open source RDBMS. It is similar to other dialects of the **SQL** databases that exist – for example, MySQL, MariaDB, and Oracle. This database can be used to store data for any web application. PostgreSQL has enterprise-grade features that make it robust, scalable, and reliable.

We will also set up SQLAlchemy, an **object-relational mapper** (**ORM**). An ORM is a high-level abstraction layer on top of a relational database that allows web developers to rely on Python object-oriented programming code to execute database operations, such as read, insert, update, and delete, rather than writing SQL queries directly. Finally, in this section, we will set up Alembic to handle database migrations.

Setting up PostgreSQL

To get started with PostgreSQL locally on your machine, download it from https://www.postgresql.org/download/ and select your operating system package.

Figure 8.4 – A diagram showing the download page of PostgreSQL

Run through the installation wizard by following the instructions to set up the database. During the PostgreSQL installation, you will be prompted to enter a super user password. It is very important you keep the password you entered as super user, as this will be required to log in to the default PostgreSQL database. Once the installation is complete, log in to the database.

Open the command terminal on your machine and type the following command:

```
$    psql -d postgres -U postgres
```

`psql` invokes a connection to PostgreSQL from the terminal. Then, you can use the `-d` option to select the database you want to access and `-U` to select the user with access permission to the database.

> **Note**
> If the command terminal replies with `psql not recognized as an internal or external command`, you might need to add the Postgres `bin/` (`C:\Program Files\PostgreSQL\14\bin`) to the system path of the environment variable.

Then, enter the password you created for the superuser during the installation.

The following screenshot shows the terminal while trying to log in to PostgreSQL.

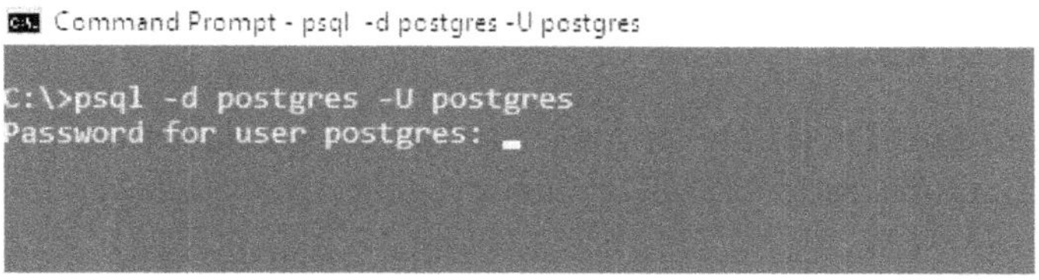

Figure 8.5 – A screenshot showing how to access PostgreSQL from the terminal

It is always a best practice to create a new user on Postgres, different from the superuser created during the installation. Now that you are logged in with the default Postgres user, let's create another user role called `admin` and assign it the password `admin123`:

```
CREATE ROLE admin WITH LOGIN PASSWORD 'admin123';
```

We will now give a new user `admin` permission to create a database:

```
ALTER ROLE admin CREATEDB;
```

With all this set, log out from the Postgres user and log in as `admin`. To log out, type the `\q` command.

Now, log in as an `admin` user:

```
$    psql -d postgres -U admin
```

Enter the admin password and log in. Once logged in, create a new database called `bizza`:

```
$    CREATE DATABASE bizza;
```

To connect to the newly created database, run the following command:

```
$    \connect bizza
```

The following screenshot shows how to connect to create a database in PostgreSQL:

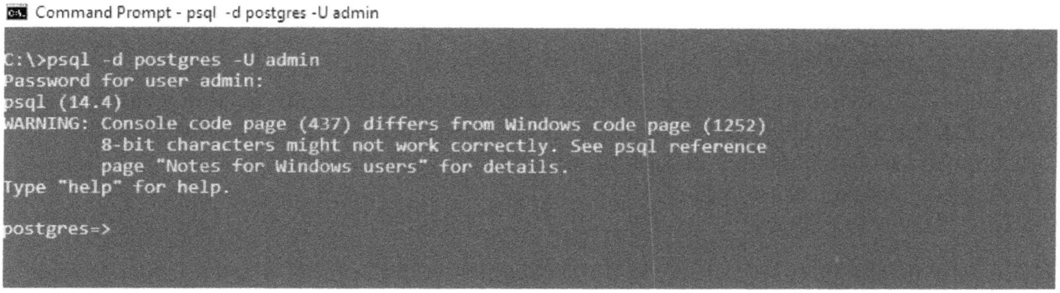

```
Command Prompt - psql  -d postgres -U admin
C:\>psql -d postgres -U admin
Password for user admin:
psql (14.4)
WARNING: Console code page (437) differs from Windows code page (1252)
        8-bit characters might not work correctly. See psql reference
        page "Notes for Windows users" for details.
Type "help" for help.

postgres=>
```

Figure 8.6 – A screenshot showing a connection to the bizza database

Hurray! We have set up PostgreSQL successfully. Now, let's dive into setting up SQLAlchemy.

> **Note**
>
> The role name, `admin`, and password, `admin123`, could be anything you want; they don't have to be named as suggested.

Setting up SQLAlchemy

Setting up SQLAlchemy for a Flask application is simple and straightforward. To set up SQLAlchemy in your project, let's install SQLAlchemy with the following command in the Terminal right inside your project `root` directory. Make sure your virtual environment is activated before running the installation command:

```
pip install SQLAlchemy
```

> **Note**
>
> SQLAlchemy is a Python SQL toolkit and ORM library that provides a set of high-level APIs to interact with relational databases.

In addition, we will use a Flask extension called **Flask-SQLAlchemy**. Let's install it by running the following command in the terminal:

```
pip install flask-sqlalchemy
```

> **Note**
>
> The Flask-SQLAlchemy extension provides a wrapper for SQLAlchemy, making it easier to use SQLAlchemy in a Flask application. Flask-SQLAlchemy provides additional features, such as automatic session handling, integration with Flask's application context, and support for Flask-specific configuration options.

SQLAlchemy will be discussed deeply in the subsequent section of this chapter, *Understanding SQLAlchemy ORM basics*.

Briefly, let's demonstrate how you can seamlessly integrate SQLAlchemy into your Flask application. The process of integrating SQLAlchemy is as follows:

1. Create a new directory and name it `sqlalchemy`.

2. Then, follow the Flask project setup in the *Setting up the development environment with Flask* section of *Chapter 1*.

3. Make sure your development virtual environment is activated.

4. In the terminal, run the following commands:

    ```
    pip install flask
    ```

    ```
    pip install flask-sqlalchemy
    ```

5. Create a file named `app.py` within the `sqlalchemy` directory, and add the following code snippet:

    ```
    from flask import Flask, render_template
    from flask_sqlalchemy import SQLAlchemy
    app = Flask(__name__, template_folder='templates')

    app.config["SQLALCHEMY_DATABASE_URI"] =
        "sqlite:///bizza.db"
    db = SQLAlchemy(app)
    ```

```
@app.route('/users')
def get_users():
    users = User.query.all()
    return render_template('users.html', users=users)

class User(db.Model):
__tablename__ = "users"
    id = db.Column(db.Integer, primary_key=True)
    username = db.Column(db.String(55), unique=True)
    name = db.Column(db.String(55), unique=False)
    email = db.Column(db.String(100), unique=True)

    def __repr__(self):
        return '<User {}>'.format(self.username)
```

The preceding code sets up a basic Flask application that uses SQLAlchemy for database operations. Let's explain what happens in the preceding code:

- `from flask import Flask`: This imports the `Flask` class from the Flask package.

- `from flask_sqlalchemy import SQLAlchemy`: This imports the `SQLAlchemy` class from the Flask-SQLAlchemy package.

- `app = Flask(__name__, template_folder="templates")`: This creates a new `Flask` application instance. `template_folder` is added to ensure that Flask finds the `templates` directory. `App.config["SQLALCHEMY_DATABASE_URI"]` `= "sqlite:///bizza.db"`: This sets the database URI for the SQLite database. In this case, the database file is called `bizza.db` and is located in the `instance` directory of the Flask application file, `app.py`. We use `SQLite` in this example. You can use any database dialect – MySQL, Oracle, PostgreSQL, and so on.

- `db = SQLAlchemy(app)`: This creates a new `SQLAlchemy` instance that is bound to the Flask application.

- Then, we have a Flask route handler that responds to the `/users` URL endpoint. When a user visits the `/users` URL, Flask will execute the `get_users` function, retrieve all the users from the database, and render the `users.html` template, passing the user's variable to the template for display.

- Lastly, we have the `User` model with four columns – `id`, `username`, `name`, and `email`. Each column represents a field in the corresponding database table and specifies its data type and optional constraints, such as uniqueness. This represents how we use SQLAlchemy in Flask to define our models as Python classes, specify their attributes and relationships, and perform database operations.

Now, let's create the `users` table in the terminal of the `sqlalchemy` directory.

6. Enter the following commands in the terminal:

```
flask shell
from app import db
db.create_all()
```

The preceding commands create the users table. You can load the `users` table with sample data with python `sample_data.py`. The `sample_data.py` is located inside the project root directory in GitHub.

7. Then, enter the `flask run` command and go to `http://127.0.0.1:5000/users` to view the users list from the database.

The following is a screenshot that shows how you can use SQLAlchemy to define and retrieve users' data.

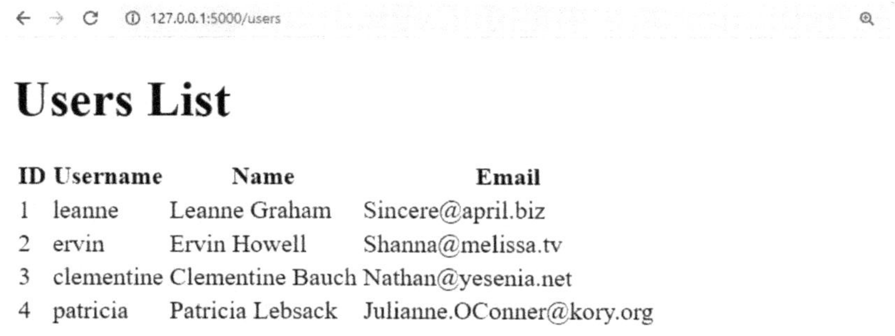

Figure 8.7 – A screenshot showing how to use SQLAlchemy for database operations

With this setup, we have demonstrated how you can define the SQLAlchemy model and use Python objects to interact with the database in a Flask application.

Next, we will dive into how you can set up Alembic, a database migration tool.

Setting up Alembic

Alembic is simply a database migration tool. You can install it with `pip install alembic` or install it from another migration package, `Flask-Migrate`, using `pip install flask-migrate`. `Flask-Migrate` relies on Alembic for database migration, and we will use it here as a tool of choice for database migration.

The `Flask-Migrate` migration tool allows you to track changes to your database. Migration tools come with the ability to manage database structure and operations using a migration script. The essence of Alembic, a migration tool, is to facilitate the automatic generation of migration SQL scripts.

In the last section of this chapter, we will discuss migration in depth with some implementation details. Next, we will discuss how you can interact with a database from a Flask web application.

Understanding database concepts for Flask applications

Now that we have set up our database and connected with it using the Terminal, it is crucial to have a solid understanding of some database concepts to be able to set up backend services that can collect, store, and retrieve users' data.

Most modern web applications have a database to store users' data. As a full stack web developer, part of your responsibility is to be able to set up backend services that can collect, store, and retrieve users' data. We will dive into interacting with databases from a Flask application shortly, but before that there are few database concepts you need to understand.

Let's take a quick overview of the following database concepts.

RDBMS

When working with a database in a production environment, you need an RDBMS. An RDBMS is a software package that allows you to interact with a database. The RDBMS software allows you to define, manipulate, retrieve, and manage data in a database.

You have database features that allow you to manage data in a database and the structure of the database itself. There are many different RDBMS flavors on the market – MySQL, PostgreSQL, Oracle, MsSQL, and MariaDB – but the one we'll work with is called PostgreSQL.

Let's take a look at another relevant tool needed to interact with a database, called a **database application programming interface (DB-API)**.

DB-APIs

A DB-API is an **application programming interface (API)** that allows communication between a programming language or web server framework and a database server, using protocols such as TCP/IP. Python, a language we use with the Flask framework, uses a library called psycopg2 that allows Python modules to interact with a PostgreSQL database server.

The DB-API acts as a database adapter, a web interface that allows a web server programming language – Python, in our case – to run SQL queries on a database, using psycopg2 as a library for the DB-API.

Different types of DB-APIs exist for every programming language or server framework and for the various dialects of the SQL database that we have. For instance, we have psycopg2 for PostgreSQL, `mysql-python` or `PyMySQL` for MySQL, `adodbapi` or `pymssql` for MS SQL Server, and `mxODBC` or `pyodb` for Oracle.

The documentation for PEP 248 DB-API version 1.0 specifies a goal for the DB-API:

> *"This API has been defined to encourage similarity between the Python modules that are used to access databases. By doing this, we hope to achieve a consistency leading to more easily understood modules, code that is generally more portable across databases, and a broader reach of database connectivity from Python."* (`http://www.python.org/dev/peps/pep-0248/`)

In essence, the DB-API gives you a low-level way of writing SQL query statements, making interactions with different databases consistently simple and easy.

Client–server model interaction

Client-server model interaction is a communication paradigm in which a client requests resources or services from a server over a network. In this model, the client initiates a request to the server, and the server provides the requested service or data in response, creating a client-server interaction that forms the foundation of various applications and network communications.

In the client-server model interaction, the browser acts as a client and the web server as a server. When an end user requests a resource or web page from the server, the request is made through a browser (the client side) through HTTP protocols on a web server (the server side). This same client-server architecture can also be modeled on an end user requesting a resource or web page that contains data from a database.

When a browser makes a request that requires data from the database, the web server receives the request and initiates a connection with the database server. The web server becomes the client, and the database server becomes the server, completing a client-server model from the backend infrastructure (`http://www.python.org/dev/peps/pep-0248/`).

Later in this chapter, we will create tables to thoroughly understand how to interact with a database. The following diagram shows the client–server interaction.

Figure 8.8 – A diagram showing the client–server interaction

In the preceding diagram, the client initiates the interaction by sending a request to the server. The request can be an HTTP request in the case of web-based applications. The server receives the request, interprets it, and performs the necessary operations to generate a response.

Once the response is ready, the server sends it back to the client. The client receives the response and processes it, utilizing the information provided to complete the client-server interaction. Next, we will dive into ORM basics to understand how you can use Python classes to interact with an RDBMS.

Understanding SQLAlchemy ORM basics

SQLAlchemy offers developers the ability to work entirely in Python code to create, read, update, and delete tables. SQLAlchemy is the Python SQL toolkit and ORM that allows application developers to interact with databases without writing direct SQL statements.

As SQLAlchemy is an ORM library, Python classes are mapped to tables, and the instances of those classes are mapped to table rows in a relational database. You then have a situation where you can use Python object-oriented programming code to perform database SQL **create**, **read**, **update**, and **delete** (**CRUD**) operations and other necessary operations in your applications.

The ORM feature of SQLAlchemy gives Python developers the power to harness function calls to generate SQL statements out of the box. With this new way of thinking about databases, developers are able to decouple object models and database schema, leading to a more flexible database structure, elegant SQL statements, and Python code that can interact with different types of database dialects.

SQLAlchemy doesn't mind the kind of database systems you work with. SQLAlchemy makes it easier to switch from one database to another without altering your code. SQLAlchemy is undoubtedly a powerful tool that simplifies the process of web development, making it faster and more efficient.

With SQLAlchemy's user-friendly interface, developers can easily create database models and interact with them from their Flask application. Additionally, SQLAlchemy is designed to reduce the chances of introducing bugs into the code base, which is especially important for large-scale applications.

With the provision of a robust framework to handle database interactions, SQLAlchemy allows developers to focus on building the logic of their applications without worrying about the underlying database operations. This results in a more streamlined development process and better overall code quality.

Lastly, the SQLAlchemy ORM library comes with automatic caching. SQLAlchemy caches collections and references between objects once initially loaded. This invariably improves performance and prevents you from sending SQL queries to the database upon every call.

In SQLAlchemy, there are three main vital layers of abstraction. These layers are the engine, the dialect, and the connection pool. This trio describes how you choose to interact with a database.

Figure 8.9 – The SQLAlchemy layers of abstraction

Let's dig deeper into the SQLAlchemy layers of abstraction to better understand how they are used in SQLAlchemy.

The SQLAlchemy engine

An SQLAlchemy application starts with the creation of an engine. You need to create an engine before SQLAlchemy can connect and interact with a database. The engine is at the lowest layer of abstraction in SQLAlchemy, and it works in pretty much the same way in which we use Psycopg2 to interact with a database.

Let's create an engine to kickstart database connectivity and interaction.

The following snippet demonstrates PostgreSQL database connectivity using `create_engine` from SQLAlchemy. The SQLAlchemy engine connects to a PostgreSQL database and executes a SQL query in Python:

```
from sqlalchemy import create_engine
# create the engine with the database URL
engine = create_engine
    ("postgresql://admin:admin123@localhost:5432/bizza")
# create a connection to the database
conn = engine.connect()
# execute a SQL query
result = conn.execute('SELECT * FROM speakers')
# loop through the result set and print the values
for row in result:
    print(row)
# close the result set and connection
result.close()
conn.close()
```

This is what happens in the preceding code:

- The `create_engine` function is imported from the `sqlalchemy` library:

```
from sqlalchemy import create_engine
```

- Then, the `create_engine()` method is invoked in SQLAlchemy and assigned to the engine instance. You then pass in a connection URL string that specifies backend server details, such as the database's name, `username`, `password`, `host`, and `port` used for connectivity:

```
engine = create_engine(
    'postgresql://admin:admin123@localhost:5432/bizza'
    )
Database name-bizza
Username-admin
Passwordadmin123
```

- A connection pool is created for the current database, and this is created once in the lifetime application:

```
connection = engine.connect()
```

- We execute the SQL statements using the established connection object:

```
result= connection.execute('SELECT * FROM speakers')
```

- Finally, we loop through the result set returned by the query and print the values.

It's important to close the result set and the connection using the `result.close()` and `conn.close()` methods to free up resources and prevent memory leaks.

SQLAlchemy connection pools

Connection pooling is an implementation of the object pool design paradigm in software engineering, where connections are reused instead of being created every time a connection to a database is required. In SQLAlchemy, the use of the `create_engine()` method usually generates a connection pool object that is created once and reused in subsequent connections when carrying out database transactions.

This connection pooling design pattern helps improve performance and better manage application concurrent connections. The default setting of the connection pool can be adjusted to better serve end users of business web applications efficiently.

To update the connection pool, add pooling parameters to `create_engine()`:

```
engine = create_engine('postgresql://user:admin123@localhost/bizza',
pool_size=20, max_overflow=0)
```

The following bullet list shows you how to tweak the connection pool parameters to optimize performance while working with the SQLAlchemy connection pool.

- With `pool_size`, you can set the number of connections the pool can handle

- With `max_overflow`, you can specify how many overflow connections the pool supports

- With `pool_recycle`, you can configure the maximum age in seconds of connections in the pool

- With `pool_timeout`, you can specify how many seconds the application needs to wait before giving up getting a connection from the pool

SQLAlchemy dialect

One of the main benefits of using SQLAlchemy is that you can use different types of DB-API implementations (Psycopg2) and relational databases without altering your Python code to suit their internal workings.

So, the dialect is a system in SQLAlchemy that makes interaction with different databases possible. You can use SQLite in your development environment and decide to use MySQL or PostgreSQL in production with your existing code base.

Some examples of relational database dialects are as follows:

- PostgreSQL

- MySQL and MariaDB

- SQLite

- Oracle

- Microsoft SQL Server

In order to use these database dialects, you must install appropriate DB-API drivers in your application. When connecting with PostgreSQL, SQLAlchemy uses Psycopg2 as an implementation of the DB-API specification. Psycopg2 is a PostgreSQL adapter for Python. Psycopg2 provides a simple and efficient way to communicate with a PostgreSQL database using Python code.

We've talked about how SQLAlchemy dialects work, but what about the data types that are used within those dialects? SQLAlchemy offers a wide variety of data types, and we'll explore some of them now. After that, we'll discuss how to map them to Python classes.

SQLAlchemy data types – mapping between tables and classes

SQLAlchemy provides high-level abstractions to any underlying relational databases we may choose to use in our applications. SQLAlchemy has data type support for the most common traditional databases we are familiar with. For instance, dates, times, strings, integers, and Booleans are well supported by SQLAlchemy.

SQLAlchemy data types allow for precise data storage and retrieval. Each column in a database table has a specific data type that defines the type of data that can be stored in that column, such as integers, strings, or dates. SQLAlchemy provides a range of data types that can be used to define the columns in a database table, including Numeric, String, DateTime, and Boolean.

By using these data types, we can ensure that our data is stored accurately and efficiently, allowing for faster and more reliable data retrieval. Also, the mapping between tables and classes is important because it allows us to work with database tables in an object-oriented manner.

SQLAlchemy provides an ORM system that allows us to map database tables to Python classes. This means that we can work with database data using Python objects and methods, which makes it easier to build and maintain our web applications. Let's see an example of how SQLAlchemy data types are mapped with Python classes and instance attributes for table and column creations, respectively.

This approach will be used to define tables from Python classes for the Flask applications in this book:

```
class Speaker(Base):
    __tablename__ = 'speakers'
speaker_ id=Column(Integer, primary_key=True)
first_name=Column(String)
last_name=Column(String)
email=Column(String)
created_at = Column(DateTime(timezone=True),
    server_default=datetime.now)
updated_at = ColumnDateTime(timezone=True),
    default=datetime.now, onupdate=datetime.now)
```

In the preceding code snippet, we defined a class called Speaker that has some attributes. Let's dive deeper into the code:

- The __tablename__ attribute is set to speakers, indicating that instances of the Speaker class should be stored in the speakers table in the database.
- The speaker_id attribute specifies that this is the primary_key in the table and has an Integer type.
- The first_name attribute specifies that a column in the table has the name first_name with a String type.
- The last_name attribute specifies that a column in the table has a name last_name with a String type.
- The email attribute specifies that a column in the table has the name email with a String type.
- created_at and updated_at specify columns in the table with the mentioned names that are of type date. Methods are, however, passed to get the current timezone.

With the classes and instances defined, we can utilize the internal APIs of SQLAlchemy and the configured dialect to map the class attributes to the corresponding native structure of the database. For example, if we have a string data type in our class, SQLAlchemy will map it to a `varchar` column in PostgreSQL.

This ensures that the data types used in the class are properly translated into the appropriate database column types, allowing for seamless communication between the Python code and the database.

To build a robust web application for a speakers' conference web application, we need to model the data correctly. SQLAlchemy provides us with a powerful ORM that makes it easy to map our database tables to classes.

We can then define the attributes and relationships between these classes, which makes it easy to work with complex data structures. In the next section, we will explore how to model data for our speakers' conference web application using SQLAlchemy.

Modeling data for a speakers' conference web application

Data modeling is the process of creating a conceptual, logical, and visual representation of data structures and relationships between data elements, in order to understand, analyze, and manage complex information systems. Data modeling involves identifying entities (objects, concepts, or things) that will be represented in a system and defining their attributes and relationships with other entities.

The main purpose of data modeling is to create a clear and precise representation of the data that will be stored, processed, and managed by an information system. A well-designed data model in a web application can ensure that the web application is scalable, efficient, and meets the needs of its users.

Let's quickly examine some of the best practices to be considered when designing a data model for the conference speakers' web application:

- **Identifying the entities**: Start by identifying the entities or objects that will be represented in the web application based on the system requirements, such as users, speakers, presentation, schedule, attendees, sessions, venue, and sponsors.

- **Defining the relationships**: Determine the relationships between these entities, such as one-to-one, one-to-many, or many-to-many relationships. For instance, a speaker can give multiple presentations at the conference, but a presentation can only be given by one speaker.

- **Determining the attributes**: Define the attributes or properties of each entity – for instance, a speaker could have attributes such as name, contact address, biography, or presentation topics.

- **Considering performance and scalability**: Design the data model to optimize performance and scalability, such as using indexes, caching, or denormalization to reduce query times and improve system response.

- **Planning for data storage**: Consider the data storage options available, such as relational databases, NoSQL databases, or flat files, and choose the appropriate option based on the needs of the web application. In this case, PostgreSQL was chosen.

- **Data security**: Design the data model to ensure data security, such as using encryption or access controls to protect sensitive data. For instance, when a user logs in, their entered password can be hashed and compared to the stored hash to verify their identity.

- **Considering future changes**: Plan for future changes and updates to the web application, such as adding new entities or attributes, or modifying existing relationships.

Are you prepared to begin developing the backend of the *Bizza* project by establishing and organizing the data model? The *Bizza* project aims to create a full stack web application that is database-driven and focuses on conference speakers.

In this data model, we will take a closer look at the database, tables, and relationship between the tables.

The `bizza` database has the following tables and relationships:

- **User model**: This model will include data about each user of the application, such as their username, password (encrypted), email address, and any other relevant details. This model will be used for authentication and authorization purposes. The following relationships exist with other tables – a one-to-one relationship between `users` and `userExtras`, a many-to-many relationship between users and roles, a one-to-one relationship between `users` and `speakers`, and a one-to-many relationship between `users` and `attendees`.

- **UserExtras model**: This model will include additional data about each user, such as their name, contact information, profile photo, and any other relevant details. The following relationship exists with the `users` table – a one-to-one relationship between `users` and `userExtras`.

- **Role model**: This model will include data about each role available in the system, such as administrator, speaker, attendee, or sponsor. Each role will have its own set of permissions and access controls. The following relationship exists with the `users` table – a many-to-many relationship between `users` and `roles`.

- **Speaker model**: This model will include data about each speaker, such as their name, bio, photo, contact information, and any other relevant details. The `Speaker` model will be linked to the `users` model to associate each speaker with their user account. The following relationships exist with other tables – a one-to-many relationship between `speakers` and `sessions`, a one-to-one relationship between `users` and `speakers`, a one-to-many relationship between `speakers` and `schedules`, and a one-to-many relationship between `speakers` and `presentations`.

- **Presentation model**: This model will include data about each presentation or talk, such as the title, description, date and time, duration, location, and speaker(s). The `Presentation` model will be linked to the `Speaker` model to associate each presentation with its speaker. The following relationships exist with other tables – many-to-one relationships between `presentations` and `speakers`, and one-to-many relationships between `presentations` and `schedules`.

- **Schedule model**: This model will include data about the overall conference schedule, including the dates, times, and locations of all presentations, workshops, and other events. The `Schedule` model will be linked to the `Presentation` model to associate each presentation with its date and time. The following relationships exist with other tables – many-to-one relationships between `schedules` and `presentations`, one-to-many relationships between `schedules` and `sessions`, many-to-one relationships between `schedules` and `venues`, and many-to-many relationships between `schedules` and `attendees`.

- **Attendee model**: This model will include data about each attendee, such as their name, contact information, registration status, and any other relevant details. The following relationship exists with other tables – many-to-many relationships between `attendees` and `schedules`.

- **Session model**: This model will include data about each session, which can be a group of related presentations or workshops. The `Session` model will be linked to the `Presentation` model to associate each session with its presentations. The following relationships exist with other tables – a many-to-one relationship between `sessions` and `schedules`, and a one-to-one relationship between `sessions` and `presentations`.

- **Venue model**: This model will include data about the conference venue, such as the address, capacity, layout, and any other relevant details. There is a one-to-many relationship between `venues` and `schedules`.

- **Sponsor model**: This model will include data about each sponsor of the conference, such as their name, logo, website, and any other relevant details.

Now that we have defined our data model, we will explore in the next section how to seamlessly send data to a PostgreSQL database with just a few simple steps.

Sending data to the PostgreSQL database from a Flask app

Interacting with a database is a common aspect of most web applications. Sending data to the PostgreSQL database from a Flask app is simple and straightforward. A web application would not be complete without the ability to store and retrieve data from a database.

The first step in sending data to a database is to ensure there is a connection between the Flask app and the database. This involves installing the required libraries and ensuring that you work in a virtual environment to contain your installation and prevent unexpected occurrences, due to the interference of other libraries.

Let's make use of the `bizza/backend` directory created in the *Setting up the development environment with Flask* section of *Chapter 1*. You can create one if you haven't already. Install and activate the virtual environment.

To create a virtual environment, open your project root directory in a terminal and add the following command:

```
python -m venv venv
```

Now, let's activate the virtual environment:

- **For Windows users**:

    ```
    Venv\Scripts\activate
    ```

- **For Linux/Mac users**:

    ```
    source venv/bin/activate
    ```

To create a database table from a Python class, you need `flask-sqlalchemy` installed. We have this installed already. If not, enter the following:

```
pip install flask-sqlalchemy
```

In the Flask application module – that is, `app.py` – update the content of the `app.py` with the following code that creates the `users` table in the *bizza* database. In this section, we will only create the `users` table to demonstrate how to send data from a Flask app to PostgreSQL.

The other models will be shown on GitHub and later in the book until we have a full implementation of the speakers' conference web app, *Bizza*:

```python
# Import flask
from flask import Flask
# Import datetime
from datetime import datetime
# Import SQLAlchemy
from flask_sqlalchemy import SQLAlchemy
# Create a Flask instance
app = Flask(__name__)
# Add the PostgreSQL database
app.config['SQLALCHEMY_DATABASE_URI'] =
    'postgresql://<db_username>:<db_password>@localhost:
    5432/<database_name>'

# Initialize the database
db = SQLAlchemy(app)

# User model definition
class User(db.Model):
```

```
    __tablename__ = 'users'

    user_id = db.Column(db.Integer, primary_key=True)
    username = db.Column(db.String(100), unique=True,
        nullable=False)
    email = db.Column(db.String(120), unique=True,
        nullable=False)
    password = db.Column(db.String(128), nullable=False)
    first_name = db.Column(db.String(100), nullable=False)
    last_name = db.Column(db.String(100), nullable=False)
    roles = db.Column(db.String(100))
    is_active = db.Column(db.Boolean, default=True)
    is_superuser = db.Column(db.Boolean, default=False)
    created_at = db.Column(db.DateTime,
        default=datetime.utcnow())
    updated_at = db.Column(db.DateTime,
        default=datetime.utcnow,
        onupdate=datetime.utcnow())


    def __repr__(self):
        return '<User %r>' % self.username
```

Let's delve into the preceding Flask app snippet used to create the `users` table in the database:

- `from flask import Flask` imports the `Flask` class from the Flask module.

- `from flask_sqlalchemy import SQLAlchemy` imports the `SQLAlchemy` class from the `flask_sqlalchemy` module.

- `app = Flask(__name__)` creates a `Flask` instance named app.

- `app.config['SQLALCHEMY_DATABASE_URI'] = 'postgresql://<db_username>:<db_password>@localhost:5432/<database_name>'` is where the `app.config[]` dictionary defines the path to the database with the `db username` and `password` set.

- `db = SQLAlchemy(app)` initializes an instance of the `SQLAlchemy` class with the Flask app as the argument.

- Then, we start to define the `model` class for `User` with `class User(db.Model):`. The `User(db.Model)` class defines a `User` model that inherits from the `db.Model` class.

- `__tablename__ = 'users'` allows us to have a custom name for the table – a *Class User* model with a corresponding table name, `users`. If this is not specified, the lowercase class name (`user`) will be used.

- Then, we add the columns for the table. Each column is an object of the `Column` subclass of SQLAlchemy:

  - `User_id = db.Column(db.Integer, primary_key=True,nullable=False)` defines a primary key column named `user_id` with an `Integer` data type

  - `username = db.Column(db.String(50), unique=True, nullable=False)` defines a column named `username` with a `String` data type and enforces that it must be unique and not nullable

  - `email = db.Column(db.String(120), unique=True, nullable=False)` defines a column named `email` with a `String` data type and enforces that it must be unique and not nullable

  - `password = db.Column(db.String(256), nullable=False)` defines a column named `password` with a `String` data type and enforces that it must not be nullable

  - `first_name = db.Column(db.String(50), nullable=False)` defines a column named `first_name` with a `String` data type and enforces that it must not be nullable

  - `last_name = db.Column(db.String(50), nullable=False)` defines a column named `last_name` with a `String` data type and enforces that it must not be nullable

  - `is_active = db.Column(db.Boolean, default=True)` defines a column named `is_active` with a `boolean` data type and sets a default value of `True`

  - `is_superuser = db.Column(db.Boolean, default=False)` defines a column named `is_superuser` with a `Boolean` data type and sets a default value of `False`

- `def __repr__(self)` defines a method that returns a string representation of the `User` object.

- `return '<User %r>' % self.username` returns a string that includes the username of the `User` object.

Now that we have defined the `User` class model, it's time to create the `users` table in the database and pass in some data:

1. Open a terminal in the project root directory and enter the `flask shell` command.

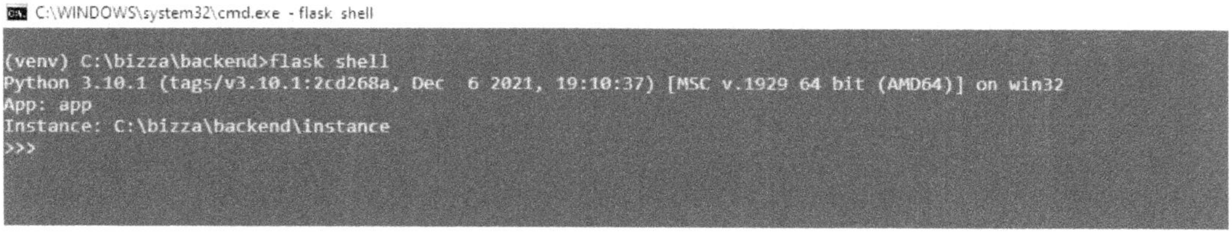

C:\WINDOWS\system32\cmd.exe - flask shell

```
(venv) C:\bizza\backend>flask shell
Python 3.10.1 (tags/v3.10.1:2cd268a, Dec  6 2021, 19:10:37) [MSC v.1929 64 bit (AMD64)] on win32
App: app
Instance: C:\bizza\backend\instance
>>>
```

Figure 8.10 – A screenshot showing flask shell

2.   Enter `from app import db` to connect with the database object.

3.   Enter `db.create_all()` to create the `users` table.

Select C:\WINDOWS\system32\cmd.exe - flask shell

```
(venv) C:\bizza\backend>flask shell
Python 3.10.1 (tags/v3.10.1:2cd268a, Dec  6 2021, 19:10:37) [MSC v.1929 64 bit (AMD64)] on win32
App: app
Instance: C:\bizza\backend\instance
>>> from app import db
>>> db.create_all()
```

Figure 8.11 – A screenshot showing db.create_all()

4.   Enter `from app import User` to gain access to the `User` model.

5.   Add user data to the table using the `flask shell` terminal:

```
user1=User(username='john',first_name='John',last_
name='Stewart',email='admin@admin.com',password='password')
```

6.   Add `db.session.add(user1)` and `db.session.commit()` to add a user to the database session and a commit for data persistence in the database.

7.   Enter `User.query.all()` to see the `users` table with the added information.

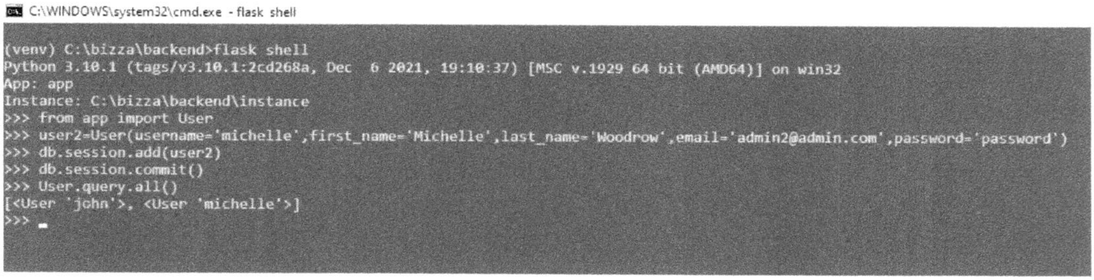

C:\WINDOWS\system32\cmd.exe - flask shell

```
(venv) C:\bizza\backend>flask shell
Python 3.10.1 (tags/v3.10.1:2cd268a, Dec  6 2021, 19:10:37) [MSC v.1929 64 bit (AMD64)] on win32
App: app
Instance: C:\bizza\backend\instance
>>> from app import User
>>> user2=User(username='michelle',first_name='Michelle',last_name='Woodrow',email='admin2@admin.com',password='password')
>>> db.session.add(user2)
>>> db.session.commit()
>>> User.query.all()
[<User 'john'>, <User 'michelle'>]
>>>
```

Figure 8.12 – A screenshot showing flask shell with inserted data

Next, we will discuss how you can add and track changes to your database structure with migrations.

Migration with Alembic

As discussed earlier, Alembic is a migration tool that makes the tracking of changes in a database a less problematic operation for Flask developers. Since we expect data models to change, we need a tool that can keep track of these changes and ensure they are updated in the database.

This is similar to how we do version control of source code using Git. The same applies to database schema change management, where we keep incremental and reversible changes in the database structure. Working with a database table, you will want to add or remove columns, thus altering the schema in a Python model class.

Once this is done, you need an automatic process to ensure your database table and the state of the data model schema are in sync. Alembic graciously handles the schema migration and ensures that the data model in a Python file is the same as the database structure.

Let's examine how you can implement migration in a Flask application. We will add another model and use migration to track changes made to the database. Here, we add the Speaker model class.

Install Flask-Migrate with pip, as stated in the *Setting up Alembic* section:

```
pip install Flask-Migrate
```

Inside app.py, add the following snippet:

```
from flask import Flask
from flask_sqlalchemy import SQLAlchemy
from flask_migrate import Migrate
app = Flask(__name__)
app.config['SQLALCHEMY_DATABASE_URI'] =
    'sqlite:///bizza.db'
db = SQLAlchemy(app)
migrate = Migrate(app, db)

class Speaker(db.Model):
    id = db.Column(db.Integer, primary_key=True)
    name = db.Column(db.String(50), nullable=False)
    bio = db.Column(db.Text, nullable=False)
    photo = db.Column(db.String(100))
    contact_info = db.Column(db.String(100))
    user_id = db.Column(db.Integer,
        db.ForeignKey('users.user_id'), nullable=False)
    user = db.relationship('User',
```

```
        backref=db.backref('speaker', uselist=False))

    def __repr__(self):
        return f"Speaker('{self.name}', '{self.bio}',
            '{self.photo}', '{self.contact_info}')"

if __name__ == "__main__":
    app.run(debug=True, host="0.0.0.0", port=5000)
```

A few changes were made to the app.py file, including the installation of the flask-migrate package, the creation of an instance of the Migrate class with the app and db instances as arguments, and the addition of a speaker model class that will be tracked upon its inclusion in the database.

Running migrations

Based on the changes made to app.py with the addition of the speaker model class, let's implement migrations with Alembic:

1. Firstly, inside the project directory, enter the pip install flask-migrate command. The flask-migrate extension provides database migration support for Flask applications.

2. To generate the initial migration in Flask, enter the following in the command terminal of the project root directory:

   flask db init

   The preceding command will create a new directory called migrations in your Flask application directory.

3. Then, once you have initialized the migration repository, you can use the flask db migrate command to generate the first migration, based on changes you have made to your models. We have added the new speaker model to the app.py file.

4. Now, use the flask db migrate -m 'first migration message, speaker model added' command to generate a new migration script, based on the changes made to the database models. The -m flag is used to specify a message describing the changes made in the migration. The following command will create a new migration file with the changes specified in the models:

   flask db init
   flask db migrate -m "first migration speaker model"

The following screenshot shows the effects of the commands:

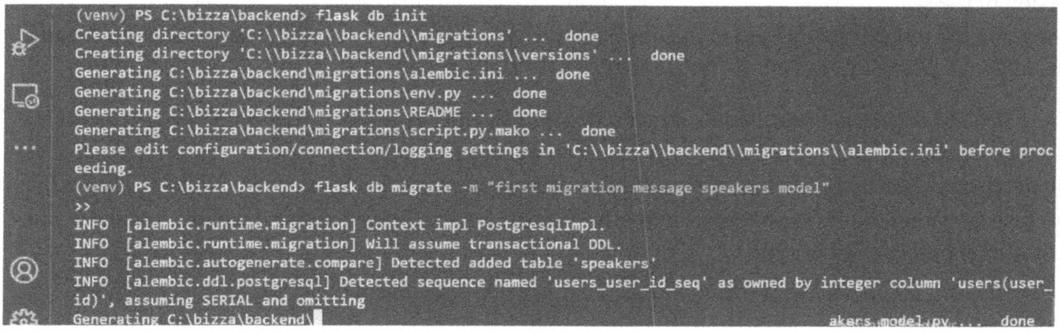

```
(venv) PS C:\bizza\backend> flask db init
Creating directory 'C:\\bizza\\backend\\migrations' ... done
Creating directory 'C:\\bizza\\backend\\migrations\\versions' ... done
Generating C:\bizza\backend\migrations\alembic.ini ... done
Generating C:\bizza\backend\migrations\env.py ... done
Generating C:\bizza\backend\migrations\README ... done
Generating C:\bizza\backend\migrations\script.py.mako ... done
Please edit configuration/connection/logging settings in 'C:\\bizza\\backend\\migrations\\alembic.ini' before proc
eeding.
(venv) PS C:\bizza\backend> flask db migrate -m "first migration message speakers model"
INFO  [alembic.runtime.migration] Context impl PostgresqlImpl.
INFO  [alembic.runtime.migration] Will assume transactional DDL.
INFO  [alembic.autogenerate.compare] Detected added table 'speakers'
INFO  [alembic.ddl.postgresql] Detected sequence named 'users_user_id_seq' as owned by integer column 'users(user_
id)', assuming SERIAL and omitting
Generating C:\bizza\backend\                                                    akers.model.py ... done
```

Figure 8.13 – A screenshot showing the migration commands

5. To commit the schema data model change to a database based on the state of the migration script, run the following command:

    flask db upgrade

    You will then get the following output:

```
(venv) PS C:\bizza\backend> flask db upgrade
INFO  [alembic.runtime.migration] Context impl PostgresqlImpl.
INFO  [alembic.runtime.migration] Will assume transactional DDL.
INFO  [alembic.runtime.migration] Running upgrade  -> 2430cf47b3d5, first migration message speakers model
```

Figure 8.14 – A screenshot showing the flask db upgrade command

This will apply all the pending migrations to your database schema. From this point on, you can continue to generate new migrations as needed using the flask db migrate command.

> **Note**
>
> To summarize the commands needed for Alembic migrations, follow these steps:
>
> **pip install flask-migrate**
>
> **flask db init**
>
> **flask db migrate -m "first migration message speakers model"**
>
> **flask db upgrade**

Alembic is a database migration tool for SQLAlchemy that helps keep the database schema in sync with the application's data model. When using Alembic, you define changes to the data model in a series of migrations, which are Python scripts that modify the database schema.

These migrations are stored in a migration directory, and each migration is associated with a specific version of the schema. When you run a migration, Alembic compares the current state of the database schema to the target state defined in the migration. It then generates a set of SQL statements to modify the database schema to match the target state.

These SQL statements are executed in a transaction, which ensures that the database schema is modified in a consistent manner. With the use of Alembic to manage database migrations, you can ensure that the database schema remains in sync with the application's data model as it evolves over time. This can help prevent data inconsistencies and other issues that can arise when the database schema and data model become out of sync.

Summary

In this chapter, we extensively discussed SQL and relational data modeling for the web. A relational database helps us with the design of a database as a group of relations. We also discussed relationships that can exist in a database, such as one-to-one, one-to-many, and many-to-many relationships, allowing us to logically group relations in the database and enforce data referential integrity.

Additionally, we examined how to set up PostgreSQL. We shed light on the basics of SQLAlchemy and its associated database adapters and how they are used in Flask application development. We discussed data model design, with the *Bizza* project as a use case. Finally, we discussed how a Flask app can communicate with the database and migration in Flask to keep track of changes in a database.

In the next chapter, we will extensively discuss the API in backend development and how you can use the Flask framework to implement API design.

9

API Development and Documentation

The **application programming interface** (**API**) is core to many technologies developers use to deal with data and facilitate communication between different systems. API-enabled digital business models are fast-growing. The need for experienced developers to build innovative enterprise solutions is equally on the rise.

The API economy is evolving into a new business model for sustainable business growth with a ton of opportunities for business owners and smart executives. If there were ever a time to be a developer, it would be now, with the plethora of public APIs and valuable commercial APIs that can make application development and deployment achievable with less effort.

Previously in this book, we discussed how databases and data modeling can be used to effectively store and retrieve application data as required. This chapter presents the opportunity to dive into the heart of backend development and leverage API technology to enable seamless communication between the various client applications and backend services.

You will learn about API design and development in Flask web applications. We will touch on common API terminologies to take your understanding of API design to a higher-than-average level. You will learn the REST API best practices and how to implement database CRUD operations in Flask and SQLAlchemy.

By the time we wrap up this chapter, you will have harnessed a better understanding of RESTful API architecture and how to design and implement a RESTful API in Flask web applications. You will have acquired an improved understanding of endpoint and payload structure to deal with data efficiently.

Eventually, you will be able to build Flask web applications that can handle HTTP requests and responses. You will be able to use Flask's SQLAlchemy extension to interact with a database and perform CRUD operations.

Finally, you will test some of the API endpoints and write clear and concise documentation for the implemented API endpoints using Postman.

In this chapter, we'll be covering the following topics:

- What is an API?

- Why use an API in web development

- Endpoint and payload anatomy

- Understanding HTTP requests/responses

- Understanding HTTP status codes

- REST API design principles

- Implementing a REST API in Flask applications

- API interaction with a database via CRUD operations

- API documentation

Technical requirements

The complete code for this chapter is available on GitHub at: `https://github.com/PacktPublishing/Full-Stack-Flask-and-React/tree/main/Chapter09`.

What is an API?

API stands for **application programming interface**. On the surface, an API seems like another piece of technical jargon coined to make learning application development difficult. This is not the case. An API's core purpose is to facilitate communication between different systems based on an agreed set of rules, methods, and protocols.

In the context of web applications, an API helps omnichannel frontend applications to communicate with backend services. The growing demand for digital services is fueling innovative ideas from business organizations to make their digital assets available through the design and implementation of APIs.

As a developer, you are going to spend a great chunk of your time developing solutions that are API-driven. Knowing how to design and implement API solutions increases your skill capital and value to your employer. Broadly speaking, there are two types of APIs: **private APIs** and **public APIs**.

Private APIs are sometimes called internal APIs. A private API describes an open architecture interface that allows developers working within an organization to have access to critical organization data. With an API, it becomes easy to automate business processes and manage information flow between various business units.

Private APIs allow businesses to develop in-house solutions efficiently with the help of existing reuseable platforms. For instance, you can broaden the scope of your frontend applications from web applications to mobile applications, leveraging the same backend services.

On the other hand, public APIs describe a standardized interface that allows developers external to an organization to have programmable access to an organization's data and services meant for public consumption. This set of interfaces allows developers to build new applications or add more functionality to their applications without reinventing the wheel. In this age, quite a huge number of public APIs are available for developers' learning purposes and are a smart way to develop innovative solutions.

The following GitHub link describes some of the public APIs you can leverage: `https://github.com/public-apis/public-apis`. Platforms such as Google, Twitter, Facebook, and Spotify allow developers to access the platform's data through APIs. With this, developers are able to build on-demand services and products.

In addition, other forms of APIs are **Simple Object Access Protocol (SOAP)**, **JavaScript Object Notation-Remote Procedure Call (JSON-RPC)**, **Extensible Markup Language-Remote Procedure Call (XML-RPC)**, and **Representational State Transfer (REST)**. These sets of rules, protocols, and specifications describe how different systems can communicate over a network. While JSON-RPC and REST can be used together, exploring this integration is not within the scope of this book.

The next part of this book will examine why APIs have emerged as a key technology for businesses and developers, and how they are transforming the way software, tools, and digital services are built and consumed.

Why use an API in web development

APIs are an integral part of modern web application development. You will rarely come across a data-driven web app without some form of API implementation. The reason why APIs are so popular is not difficult to see. APIs enable seamless integration, collaboration, and innovation by providing standardized ways to facilitate efficient resource sharing across diverse applications and systems. The following are some of the benefits of the use of APIs in web development:

- APIs allow separate systems to interact, bridging communication gaps between different components of web applications
- API-driven development enables access to third-party data and services, fostering innovative solutions and reducing development time
- APIs provide a secure and scalable means of sharing information for developers and end users
- API-centric development reduces software development time by leveraging existing APIs and avoiding reinventing the wheel
- APIs have substantial financial potential, as exemplified by the significant revenue generated by Google Maps and Twilio through API access
- In healthcare, API-driven web applications facilitate the access and management of critical health data

- APIs are valuable in the travel and tourism industry for accessing real-time flight booking information and finding the best prices

- APIs play a vital role in e-commerce by integrating payment solutions and enabling seamless transactions

- API abstraction allows developers to build secure web applications with controlled data exposure and secure architecture design

Next, we will briefly explore endpoint and payload structure to understand how to define clear and logical paths for accessing resources in API design and ensure a proper data structure for the effective communication of information between the client and server.

Endpoint and payload anatomy

Endpoints and payloads are a crucial part of any API component. Endpoints facilitate access to resources on a server through the use of well-defined routes or URLs. Endpoints usually act as the actual point at which data exchange occurs between two disparate applications in a client-server environment. Payloads allow us to send data along with a request or a response. We will discuss more on payloads in a jiffy.

Let's start by examining the structure of an endpoint and the rules guiding endpoints set up in a REST API.

Understanding the endpoint structure

Endpoint structures allow you to logically organize the resources of your application. We are going to start with a `venue` resource in exploring endpoint structure. Data is usually represented as resources in a REST API. You can define an endpoint for a `venues` collection with a `venue` resource following the use of the `collection/resource` path convention.

> **Note**
> The `venue` resource represents an object or data structure accessible via a unique URL endpoint that allows clients to retrieve, create, update, or delete information about the venue.

One of the primary goals of an API designer is to clearly model data as a resource that other developers can use in their applications.

For instance, `https://example.com:5000/api/v1/venues` is a whole path that leads to the `venue` resource on an API server.

Let's go through the structure of the path:

- `https`: Secured protocol
- `example.com`: Domain name
- `500`: Port number
- `/api/v1/venues`: Endpoint

    `/api/` represents the entry point of an API endpoint, `/v1/` represents the version number of the API, and `/venues` represents the resource

We can perform the following API operations on the endpoint based on the HTTP methods:

- `GET /api/v1/venues`: Returns a list of all the venues
- `GET /api/v1/venues/id`: Retrieves a single venue identified with `id`
- `POST /api/v1/venues/`: Creates a venue resource
- `UPDATE /api/v1/venues/id`: Updates a single venue identified with `id`
- `DELETE /api/v1/venues/id`: Deletes a single venue identified with `id`

Let's retrieve information about venues using the appropriate HTTP methods. The `/api/v1/venues` URL endpoint is used to get an overview of all available venues and their associated information from the data source. The response will be in JSON format, providing a structured representation of the venue data.

For example, let's examine a venue resource request and expected response in JSON format.

With `GET /api/v1/venues`, the expected response in JSON format will be a list of all available venues:

```
[
{
"id":1
"name": "Auditorium A"
},
{
"id":2
"name": "Auditorium B"
},
]
```

With `GET /api/v1/venues/2`, the expected response in JSON format will be a specific venue resource with `id` 2:

```
[
{
```

```
"id":2
"name": "Auditorium B"
}]
```

With POST /api/v1/venues, the expected response in JSON format will be an added venue resource with the returned id 3:

```
[
{
"id":3
"name": "Auditorium C"
}]
```

With UPDATE /api/v1/venues/3, the expected response in JSON will be an updated venue resource with id 3; the new value of the name property is now Conference Hall:

```
[
{
"id":3
"name": "Conference Hall"
}
]
```

With DELETE /api/v1/venues/3, the expected response in JSON will be a deleted resource venue with id 3:

```
[
{
"id":3
}
]
```

The preceding JSON response messages depict endpoint data representation based on the requests to the server. The /api/v1/venues RESTful API endpoints with GET will return a list of available venues, GET /api/v1/venues/2 will return a specific venue with id 2, POST /api/v1/venues will add a new venue and return its id, UPDATE /api/v1/venues/3 will update the venue with id 3 and return the updated resource, and DELETE /api/v1/venues/3 will delete the venue with id 3.

Next, we will examine some of the golden rules to adhere to while designing endpoints. With these principles, you will be able to design a more intuitive and user-friendly RESTful API that will reduce the time and effort required to develop and maintain applications that use the API.

API endpoint best practices

There are principles guiding the design of a good API endpoint and by extension API development. We will briefly examine the following golden rules for designing API endpoints your team members or other developers can relate to:

- **Use nouns as a naming convention for an endpoint**: When planning your API endpoints, they should be based on the resource instead of on the action. Resources are better named as a noun rather than an action verb. This naming style allows request methods to determine what action users can take on your resource endpoint URL.

  Also, the concise and clear naming of your endpoint will allow developers to understand clearly what your endpoint represents. For instance, in the case of the `/venues` endpoint, the noun `venues` explains what the resource is all about:

  - **Good name**: `https://example.com/api/v1/venues`
  - **Bad name**: `https://example.com/api/v1/get_all_venues`

- **Use plural nouns for collections**: API collections are recommended to have plural nouns. For consistency, you can make your resource name plural as well. In the `venues` case, you can see that we used `/venues` to describe the collection, for example, `https://example.com/api/v1/venues/2`, where `id=2`, refers to a specific resource in the collection:

  - **Good name**: `https://example.com/api/v1/venues`
  - **Bad name**: `https://example.com/api/v1/venue`

- **Adhere to the collection/resource/collection rule and never make it complex**: In this `collection/resource/collection` structure, the endpoint URL starts with the collection name, followed by the resource name, and then another collection name if applicable.

  For example, in the case of a `speaker` resource, which may have a collection of `papers`, the recommended endpoint URL would be something like `/speakers/2/papers`, where `speakers` is the collection name, 2 is the ID of a specific speaker resource, and `papers` is the collection of papers associated with this particular speaker:

  - **Good name**: `https://example.com/api/v1/speakers/2/papers`
  - **Bad name**: `https://example.com/api/v1/speakers/2/papers/8/reviews`

`https://example.com/api/v1/speakers/2/papers/8/reviews` violates the recommended structure by including another collection name, `reviews`, after `papers`. This structure implies that `reviews` is a sub-collection of `papers`, which contradicts the rule of the `collection/resource/collection` pattern. Instead, we can treat them as separate resources with their own endpoints.

Take the following example:

- **To retrieve papers associated with a speaker**: `GET /api/v1/speakers/2/papers`
- **To retrieve reviews associated with a paper**: `GET /api/v1/papers/8/reviews`

By separating the endpoints, it becomes clearer that reviews are related to papers, rather than being nested within the `papers` collection.

Next, we will explore the structure of the payload and examine its role within this context.

Understanding the payload structure

The payload contains the actual data that the API is designed to work with. In this section, you will understand the data format that is sent and received by an API. You will learn how the payload is structured, including the keys and values that are used to represent the data.

With this understanding of the payload structure, you will be able to work with more complex APIs and handle larger amounts of data. As discussed earlier, an API provides an interface for the exchange of data between web services. The data in question for interacting, communicating, or sharing is the payload.

The payload is the data of interest between various web applications that want to exchange information. Technically, this is the body of the HTTP request and response in client-server communication. In an API ecosystem, when a client makes a request, in the body of the request is the data, which essentially consists of two parts: the header/overhead and the payload.

The header is used to describe the source or destination of the data in transit. The payload comes in different flavors: JSON or XML. The payloads are recognizable with the use of curly braces, { }. We will focus on the JSON format of the payload in this book.

We are choosing JSON format because JSON is easy to read and understand, easy to parse in most programming languages, supports complex data structures, is platform-independent, and uses minimal syntax.

Let's describe a typical structure of a payload with examples.

The following is a payload that a client sends to the server (the *API request* payload):

```
POST /venues  HTTP/1.1
Host: example.com
Accept: application/json
Content-Type: application/json
Content-Length: 10

{
"id":3
```

```
"name": "Conference Hall"

}
```

Note the following in the preceding code:

- The payload is indicated with data within curly braces, and it explains the information we want to send to the /venues API endpoint using the POST HTTP method
- The "Content-Type: application/json" request header describes the JSON data type of the request body
- The client also describes the response format it expects from the server with Accept: application/json

For a server-returned payload (*OK response* payload from the server), we have the following:

```
HTTP/1.1 200 OK
Content-Type: application/json
Content-Length: 10
{
"responseType": "OK",
"data": {
"id":3
"name": "Conference Hall"
}
}
```

Note the following in the preceding snippet:

- OK and the content data within curly braces are the payloads.
- You can see that the server complied with the Content-Type: application/json the client is expecting to receive.
- The JSON payload is enclosed in curly braces, { }, and consists of two key-value pairs:

  - "responseType": "Ok": This key-value pair indicates that the API successfully processed the request and returned a response. The "responseType" key has a value of "Ok".

  - "data": { "id": 3, "name": "Conference Hall" }: This key-value pair contains the actual data being returned by the API. The "data" key has a value of an object that contains information about the venue with ID 3. In this case, the venue name is "Conference Hall".

For an error payload (the *API failed response* payload), we have the following:

```
HTTP/1.1 404 Not found
Content-Type: application/json
```

```
Content-Length: 0

{
"responseType": "Failed",
"message": "Not found"
}
```

The payloads in the preceding code are `"responseType": "Failed"` and `"message": "Not found"`.

Endpoints and payloads are essential parts of API development. You need to design API endpoints that are concise and intuitive to clearly communicate your intentions to developers who may want to interact with your API data services.

Now, we will deepen the knowledge stacks and glance through HTTP requests/responses. When building web applications, it's essential to have a good understanding of how HTTP requests and responses work.

These are the building blocks of communication between clients and servers, and knowing how to work with them is crucial for building effective and efficient web applications.

Understanding HTTP requests/responses

To successfully work with APIs, you need to have an understanding of HTTP requests/responses. So, let's unmask the structure of HTTP requests and responses.

Request line

Every HTTP request begins with the request line. This comprises the HTTP method, the requested resource, and the HTTP protocol version:

```
GET /api/v1/venues  HTTP/1.1
```

In this instance, `GET` is the HTTP method, `/api/v1/venues` is the path to the resource requested, and `HTTP 1.1` is the protocol and version used.

Let's dive deeper into HTTP methods to understand how developers use different HTTP methods to specify the type of action they want to perform when making requests to the web servers.

HTTP methods

HTTP methods indicate the action that the client intends to perform on the web server resource. Commonly used HTTP methods are the following:

- `GET`: The client requests a resource on the web server
- `POST`: The client submits data to a resource on the web server

- PUT: The client replaces a resource on the web server

- DELETE: The client deletes a resource on the web server

Let's take a glance through the request headers.

HTTP request headers

HTTP headers play a critical role in facilitating communication between the client and server during an HTTP request or response. They allow both parties to include additional information alongside the primary data being transferred, such as metadata, authentication credentials, or caching directives.

Headers act as a placeholder for payloads and provide crucial context and metadata to both the client and server. For example, they can convey information about the content type, language, encoding, and size of the data being transferred.

Additionally, headers can provide details about the client's capabilities and preferences, such as the type of browser being used or the preferred language for content delivery. HTTP request headers come immediately after the request line.

Common headers are the following:

- **Host**: `www.packtpub.com/`

  The `Host` header specifies the host of the server and indicates where the resource is requested from.

- **User-Agent**: `"Mozilla/5.0 (Windows NT 10.0; Win64; x64; rv:107.0) Gecko/20100101 Firefox/107.0"`

  The `User-Agent` header tells the web server of the application that is making the HTTP request. It usually consists of the operating system (such as Windows, Mac, or Linux), version, and application vendor.

- **Accept**: `"text/html,application/xhtml+xml,application/xml;q=0.9,image/avif,image/webp,*/*;q=0.8"`

  The `Accept` header tells the web server what type of content the client will accept as the response.

- **Accept-Language**: `en-US,en;q=0.5`

  The `Accept-Language` header indicates the language.

- **Content-type**: `text/html; charset=UTF-8`

  The `Content-type` header indicates the type of content being transmitted in the request body.

Next, we will examine the request body.

HTTP request body

In HTTP, the request body refers to the additional data that is sent along with an HTTP request message, typically in the form of a payload. Unlike HTTP headers, which provide metadata about the request or response, the request body contains the actual data that the client is sending to the server. The request body can contain various types of data, including form data, JSON, XML, binary data, or text.

For example, when submitting a web form, the data entered by the user is typically sent as part of the request body. Similarly, when uploading a file or sending an API request, the data being transmitted is often included in the request body. The format and structure of the request body depend on the content type specified in the request headers.

For instance, if the content type is set to `application/json`, the request body must be a valid JSON object. If the content type is `multipart/form-data`, the request body may include multiple parts, each containing different types of data.

The following shows an HTTP request that uses the POST method to submit data to a web server:

```
POST /users HTTP/1.1
Host: example.com

{
"key":"value",
"array":["value3","value4"]
}
```

The request includes a request body in JSON format, which contains a key-value pair and an array. The key has a value of `"value"`, and the array contains two string values, `"value3"` and `"value4"`.

The following shows an HTTP request that uses the PUT method to update data on a web server:

```
PUT /authors/1 HTTP/1.1
Host:example.com
Content-type: text/json

{"key":"value"}
```

The request includes a request body in JSON format, which contains a key-value pair. The key has a value of `"value"`, and this data is intended to update the resource at the specified endpoint.

Next, we will consider HTTP responses. HTTP responses are the server's way of communicating with the client in response to an HTTP request.

HTTP responses

Understanding the various HTTP status codes and the information included in HTTP responses is essential for building robust and effective web applications. After the web server has processed an HTTP request, it is expected to send an HTTP response to the client.

The initial line of the response contains the status, which indicates whether the request was successful or unsuccessful due to an error. This status line provides critical feedback to the client about the outcome of the request:

```
HTTP/1.1 200 OK
```

The HTTP response begins with the HTTP protocol version, followed by the status code, and a reason message. The reason message is a textual representation of the status code. In the upcoming *Understanding HTTP status codes* section, we will delve into the topic of status codes in detail.

We will now begin discussing the response headers.

HTTP response headers

In HTTP, response headers provide additional information about the response message sent by the server. While the status code in the initial line of an HTTP response provides essential information about the outcome of the request, response headers can provide additional metadata about the response, such as the content type, cache settings, and server type.

Response headers are typically used to provide the client with information that can help optimize the rendering and processing of the response, such as specifying the character encoding or content length. Response headers can also be used to control the behavior of the client, such as setting caching parameters or enabling **cross-origin resource sharing** (**CORS**) for API requests.

HTTP response headers are sent by the server in the response message, immediately following the status line. The headers consist of one or more lines, each with a header field name and a value, separated by a colon. Some common response headers include `Content-Type`, `Content-Length`, `Cache-Control`, `Server`, and `Set-Cookie`.

The following is an example of an HTTP response header:

```
Date: Sun, 27 Nov 2022 02:38:57 GMT
Server: Cloudflare
Content-Type: text/html
```

Considering the preceding code block, we have the following:

- The `Date` header specifies the date and time at which the HTTP response was generated

- The `Server` header describes the web server software used to generate the response

- The Content-Type header describes the media type of the resource returned: in this case, HTML

Next, we will discuss the HTTP response body.

HTTP response body

The response body refers to the data sent by the server in response to an HTTP request. While the response headers provide metadata about the response, the response body contains the actual data that the client requested, such as HTML, JSON, XML, or binary data.

The structure and content of the response body depend on the nature of the request and the format of the requested data. For example, a request for a web page might receive an HTML response body containing the markup and content of the page, while a request for data from an API might receive a JSON response body containing the requested data.

In HTTP, the response body may contain content in certain situations, such as when the server responds with a status code of 200, which indicates that the request was successful and that the server is returning content. In other cases, when the server responds with a status code of 204, it indicates that the request was successful but that there is no content to return, so the response body may be empty:

```
HTTP/2 200 OK
Date: Sun, 27 Nov 2022 02:38:57 GMT
Server: Cloudflare
Content-Type: text/html
<html>
    <head><title>Test</title></head>
    <body>Test HTML page.</body>
</html>
```

Having discussed HTTP requests and responses, we will now begin discussing the various commonly used HTTP status codes.

Understanding HTTP status codes

HTTP status codes are three-digit numbers sent by a server in response to an HTTP request. These codes provide feedback to the client about the outcome of the request and help identify any issues that may have occurred during the transaction. The first digit of an HTTP status code indicates the category of the responses, which could be Informational, Successful, Redirection, Client Error, or Server Error.

The common status codes you'll encounter for each category are listed as follows:

`1XX Informational`

Status Code	Description	Reason
100	This code indicates an interim response from the web server informing the client to continue the request or ignore the response if the request has already been processed	Continue

`2XX Successful`

Status Code	Description	Reason
200	This code indicates the server successfully processed the request.	OK
201	This code indicates the server successfully processed the request and a resource was created.	Created
202	This code indicates the request has been received but the processing has not yet been completed.	Accepted
204	The code indicates the server successfully processed the request but is not returning any content.	No content

`3XX Redirection`

Status Code	Description	Reason
301	This code indicates that the request and all future requests should be sent to the new location in the response header	Moved Permanently
302	This code indicates the request should be sent temporarily to the new location in the response header	Found

```
4XX Client Error
```

Status Code	Description	Reason
`400`	This code indicates that the server cannot process the request due to a perceived client error.	`Bad Request`
`401`	This code indicates that the client making the request is unauthorized and should be authenticated.	`Unauthorized`
`403`	This code indicates that the client making the request does not have the right to access the content; they are unauthorized and should obtain the right to access the resource.	`Forbidden`
`404`	This code indicates that the web server did not find the requested resource.	`Not Found`
`405`	This code indicates that the web server knows the method but the targeted resource does not support the HTTP method used.	`Method Not Allowed`

```
5XX Server Error
```

Status Code	Description	Reason
`500`	This code indicates that the web server has encountered an unexpected error while processing the request.	`Internal Server Error`
`502`	This code indicates that the web server, while acting as a gateway to get a response, received an invalid response from the application server.	`Bad Gateway`
`503`	This code indicates that the web server is unavailable to process the request.	`Service Unavailable`

Before we explore how to implement a REST API in Flask web applications, it's important to grasp the underlying principles of RESTful API design. By understanding these fundamentals, we can ensure that our API is designed in a way that is intuitive, user-friendly, and efficient. So, let's take a closer look at the key principles that underpin RESTful API design.

REST API design principles

REST API, or RESTful API, describes an API that conforms to the REST architectural style using an HTTP-based interface for network communication. An API in its simplest form defines a set of rules that disparate systems or applications need to conform to in order to exchange data.

Dr. Roy Fielding, in the year 2000, presented a dissertation (`https://www.ics.uci.edu/~fielding/pubs/dissertation/rest_arch_style.htm`) that described a novel design approach that API designers are expected to follow in building applications that can stand the test of time in addition to being secure. In order to develop a RESTful system, there are architectural constraints that are worth keeping in mind.

We will examine those REST principles, such as *client-server*, *statelessness*, *caching*, *uniform interface*, *layered system*, and *code on demand*, to conform to a REST style guide.

Client-server

The REST API architecture encourages client and server communication. The client sends a network request to the server, while the server can only send a response back to the client. RESTful APIs ensure all communications start from clients, who then wait for a response from the server.

RESTful APIs enforce the separation of concerns between the client applications and server, thus making the interaction smooth and independent. Owing to the separation of concerns, web application designs are not tightly coupled as the client and server can scale without inadvertently impacting the overall application architecture.

> **Note**
>
> Separation of concerns is a design principle that aims to separate a system into distinct, independent parts, with each part responsible for a specific task or functionality. This design principle is commonly applied in software engineering and programming, including in the design of REST APIs.

Statelessness

Conforming to REST API design constraints requires all network requests to be stateless. Statelessness means a server is not expected to remember past network requests. Technically, the statelessness of a network request encourages independent interaction between the client and server.

Every request from a client to a server is expected to contain all of the important information required to understand and fulfill the request. Statelessness invariably improves performance as the server does not need to store or remember previous requests. In addition, with a stateless state in RESTful application design, the architecture is simple to set up, scalable, and reliable.

Caching

RESTful APIs are designed with caching in mind. Caching is the process of storing frequently used data in a temporary location in order to reduce the time and resources required to access it. The caching principle in REST API ensures that network information contained within a response to a request be declared implicitly or explicitly as cacheable or non-cacheable.

For instance, if a response is cacheable, the client will reuse the cached response data for similar subsequent requests. Caching improves the efficiency of server resources and reduces bandwidth usage while decreasing the loading time of a site page.

Uniform interface

A uniform interface is another design constraint that REST API designers need to implement. The REST API architectural style states that the REST API should have a single communication protocol and a standardized data format. Regardless of the system's environment, applications and servers, a uniform interface facilitates smooth interaction.

A uniform interface encourages the ease of evolvability of each system component and provides a common framework for any client application to communicate with a REST API.

REST APIs adopt HTTP as a communication protocol for client-server interaction. With HTTP, the client sends a request in a specific format, such as JSON or XML. Let's take a look at a sample request:

```
GET https://localhost:5000/api/v1/venues
```

This REST API request contains two major components – GET and the URL:

- GET is one of the HTTP methods. The GET method specifies the action the client wants to make on the server resource. There are four commonly used HTTP requests a client uses to make a request:

  - GET: To retrieve a resource

  - POST: To create a new resource

  - PUT/PATCH: To update or edit an existing resource

  - DELETE: To delete a resource

- The URL part contains the uniform resource identifier that specifies the resource of interest. In this case, we are interested in the venues resource. So, we issued an HTTP GET request to look up the location of that resource.

Furthermore, the URL is also sometimes called an endpoint. The endpoint represents the location where the API actually interacts with the client, the point at which data exchange takes place.

The client-server interaction proceeds with the host receiving and validating the GET request. The response data is returned from the target resource (/api/v1/venues). The returned data format is often in the form of JSON or the expected response format as specified by the client. JSON allows us to have standardized structured data to display the contents of the target resource.

Layered system

Modern web applications are composed of layered architectures. The client-server system could have multiple layers of servers/services each with its own responsibility, such as load balancing, security, and caching layers. The REST API design principle encourages an implementation in which the layers of systems that may exist do not alter the natural behaviors of client-server interaction.

With this constraint, any change or modification in the inner systems/servers will have zero effect on the format of the HTTP-based request and response model. A layered system enforces a clear separation of concerns and improves scalability as the client and server are highly independent and can grow at scale.

Code on demand (optional)

Code on demand is an optional constraint in RESTful API design, which allows the server to send executable code to the client in response to a request. This code can be in the form of scripts, applets, or other executable files, and can be used to extend the functionality of the client.

The code-on-demand constraint is optional because it is not always necessary or desirable for an API to provide executable code to clients. In many cases, RESTful APIs simply provide data or resources that can be consumed by client applications without the need for executable code.

However, in some cases, code on demand can be useful for providing additional functionality to clients, such as data processing, filtering, or visualization. For example, a RESTful API for data analysis could provide executable code to the client for performing complex calculations or generating visualizations based on the data.

The REST API can send code such as JavaScript code to the client application for execution. This optional feature of code on demand allows API designers to further customize the API's functionality with the ability to increase the flexibility of the API to deliver required business solutions.

The previously mentioned design principles of REST API ensure developers are able to build solutions based on architectural styles that are widely acceptable in the software development industry.

Lastly, Dr. Roy Fielding once summarized the essence of the RESTful API design principle and the overall goal in software development as follows:

> *"REST is software design on the scale of decades: every detail is intended to promote software longevity and independent evolution.*
>
> *Many of the constraints are directly opposed to short-term efficiency. Unfortunately, people are fairly good at short-term design, and usually awful at long-term design. Most don't think they need to design past the current release. There are more than a few software methodologies that portray any long-term thinking as wrong-headed, ivory tower design (which it can be if it isn't motivated by real requirements)."*

```
https://roy.gbiv.com/untangled/2008/rest-apis-must-
                  be-hypertext-driven
```

Next, we will delve into the practical implementation of a REST API within a Flask web application.

Learning how to implement a REST API in a Flask application is a valuable skill for developers who want to build web applications that can be accessed and consumed by other applications or services.

Implementing a REST API in a Flask application

Flask, as a popular Python web framework, provides developers with a flexible and lightweight solution for building web applications. With Flask, you can easily create and manage RESTful APIs. The process of implementing a REST API in a Flask application is simple.

Implementing a REST API in a Flask application involves defining API endpoints, request handlers, and data models, and possibly connecting to a database. In this section, we are going to design a REST API service using Flask that a React frontend application can consume.

We will follow a simple process of defining the resources expected in the *Bizza* application, following which we will define the URLs (endpoints) that would be used to access the resources. In addition, we will use Flask's routing system to map each endpoint to a specific function in the application.

Each function should handle the HTTP request, interact with the database (if necessary), and return an appropriate response. Finally, we will implement data serialization to serialize and deserialize data between Python objects and JSON. This ensures that the API can establish communication with clients using a standardized format.

Next, we will begin the discussion of implementing a REST API with the concept of defining the application resources.

Defining the application resources

We will start by defining the resources needed to create a conference web application that can handle all aspects of an event, from scheduling and registration to sponsorship management. The following resources are thus defined:

- `Attendees`: The people who are attending the conference
- `Speakers`: The people who are giving presentations at the conference
- `Schedules`: The schedules of the conference, including the start and end times of each session
- `Presentations`: The speakers' areas of interest and subject matter for the conference
- `Users`: The users of the event management system, including attendees, speakers, and organizers
- `Userextras`: Additional information about the users attending the event, such as dietary preferences or accessibility needs
- `Venues`: The venues where the event or conference is being held, including information about the location, capacity, and amenities
- `Sessions`: The individual sessions or talks within the conference, including information about the speaker, topic, and time
- `Sponsors`: The organizations or companies sponsoring the event, including information about their level of sponsorship, logo, and contact details

Next, we will define the API endpoints. In order to implement a functional REST API, it is necessary to define the API endpoints clearly.

Defining the API endpoints

Defining the API endpoints is a crucial step in implementing a REST API. These endpoints allow you to perform various operations on the resources of the conference web application, such as creating, reading, updating, and deleting records. We are defining the endpoints based on the resources specified in the preceding section.

Now, the specific endpoints and HTTP methods used are listed as follows:

- `Users`:
  - `GET /users`: Retrieves a list of all users
  - `GET /users/{id}`: Retrieves information about a specific user
  - `POST /users`: Creates a new user
  - `PUT /users/{id}`: Updates information about a specific user
  - `DELETE /users/{id}`: Deletes a specific user

- Userextras:

  - GET /userextras: Retrieves a list of all userextras

  - GET /userextras/{id}: Retrieves information about a specific userextra

  - POST /userextras: Creates a new userextra

  - PUT /userextras/{id}: Updates information about a specific userextra

  - DELETE /userextras/{id}: Deletes a specific userextra

- Attendees:

  - GET /attendees: Retrieves a list of all attendees

  - GET /attendees/{id}: Retrieves information about a specific attendee

  - POST /attendees: Creates a new attendee

  - PUT /attendees/{id}: Updates information about a specific attendee

  - DELETE /attendees/{id}: Deletes a specific attendee

- Speakers:

  - GET /speakers: Retrieves a list of all speakers

  - GET /speakers/{id}: Retrieves information about a specific speaker

  - POST /speakers: Creates a new speaker

  - PUT /speakers/{id}: Updates information about a specific speaker

  - DELETE /speakers/{id}: Deletes a specific speaker

- Schedules:

  - GET /schedules: Retrieves a list of all schedules

  - GET /schedules/{id}: Retrieves information about a specific schedule

  - POST /schedules: Creates a new schedule

  - PUT /schedules/{id}: Updates information about a specific schedule

  - DELETE /schedules/{id}: Deletes a specific schedule

- Presentations:

  - GET /presentations: Retrieves a list of all presentations

  - GET /presentations/{id}: Retrieves information about a specific presentation

- POST /presentations: Creates a new presentation
- PUT /presentations/{id}: Updates information about a specific presentation
- DELETE /presentations/{id}: Deletes a specific presentation

- Venues:

  - GET /venues: Retrieves a list of all venues
  - GET /venues/{id}: Retrieves information about a specific venue
  - POST /venues: Creates a new venue
  - PUT /venues/{id}: Updates information about a specific venue
  - DELETE /venues/{id}: Deletes a specific venue

- Sessions:

  - GET /sessions: Retrieves a list of all sessions
  - GET /sessions/{id}: Retrieves information about a specific session
  - POST /sessions: Creates a new session
  - PUT /sessions/{id}: Updates information about a specific session
  - DELETE /sessions/{id}: Deletes a specific session

- Sponsors:

  - GET /sponsors: Retrieves a list of all sponsors
  - GET /sponsors/{id}: Retrieves information about a specific sponsor
  - POST /sponsors: Creates a new sponsor
  - PUT /sponsors/{id}: Updates information about a specific sponsor
  - DELETE /sponsors/{id}: Deletes a specific sponsor

Once the API endpoints have been defined, the next step is to implement them in the Flask application.

Let's start digging!

Implementing the API endpoints

Implementing the API endpoints is a critical step in developing a RESTful API. This is where all the juicy bits come together to form the heart and soul of your REST API. API endpoints define the functionality and behavior of the API, specifying the methods that can be used to access the API's resources.

In a Flask application, implementing the API endpoints involves mapping the URLs to the relevant functions, defining the HTTP methods, and writing the Flask view functions that will handle the requests and generate responses. In addition, it is necessary to specify the request and response formats to be used to communicate with the API.

In this section, we will explore the process of implementing API endpoints in a Flask application. Let's start by creating a simple endpoint that returns a text-based welcome message from the API server.

In the backend development environment, inside `bizza/backend/`, activate the virtual environment in the terminal:

- **Use the following on Windows**:

  ```
  venv/Scripts/activate
  ```

- **Use the following on Mac/Linux**:

  ```
  source ./venv/bin/activate
  ```

If you are having issues activating your virtual environment, check *Setting up the development environment with Flask* in *Chapter 1*.

Now, update `app.py` with the following code:

```
from flask import Flask
app = Flask(__name__)

@app.route("/")
def index():
    return "Welcome to Bizza REST API Server"

if __name__ == "__main__":
    app.run()
```

This is what is happening in the preceding code snippet:

- We import the `Flask` class from the `flask` package.

- We then create an instance of the `Flask` class and name it app. Then, we pass in a `__name__` variable as an argument that references the current module name. This is needed for the internal working of Flask for path discovery.

- Use a `@route()` Flask decorator to tell Flask to implement the `index()` view function when a user accesses the URL `"/"` (index URL). A decorator in Python is simply a way of adding extra functionality to functions without explicitly altering the method behavior.

- This view function returns the message `Welcome to Bizza REST API server` to the browser. So, essentially, the decorator is capable of modifying the `index()` view function to return a value in the form of an HTTP response, which can then be displayed by the client using the desired data presentation format. In this case, `text/html` was returned.

- If the conditional part of the code becomes true, that is, `app.py` is the main program, then it runs the module. In this way, Python prevents the accidental or unintentional running of an imported module.

We can test the endpoint with `curl` by entering the following command:

```
curl http://localhost:5000/
```

We get the following output:

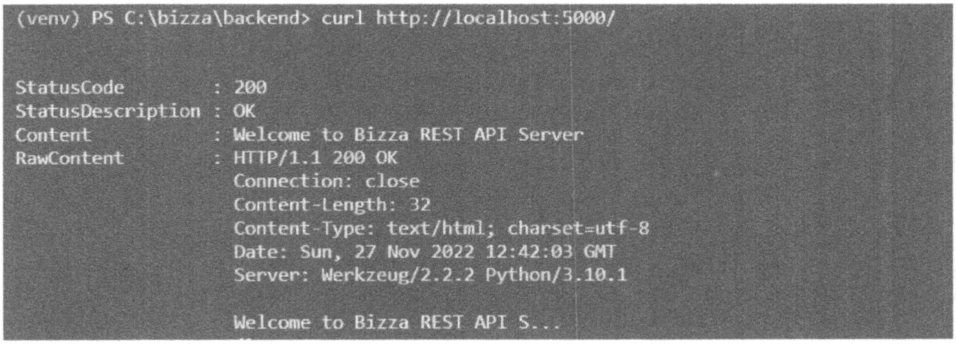

```
(venv) PS C:\bizza\backend> curl http://localhost:5000/


StatusCode        : 200
StatusDescription : OK
Content           : Welcome to Bizza REST API Server
RawContent        : HTTP/1.1 200 OK
                    Connection: close
                    Content-Length: 32
                    Content-Type: text/html; charset=utf-8
                    Date: Sun, 27 Nov 2022 12:42:03 GMT
                    Server: Werkzeug/2.2.2 Python/3.10.1

                    Welcome to Bizza REST API S...
```

Figure 9.1 – Screenshot showing the HTTP response from localhost

JSONifying response data

Jsonifying response data refers to the process of converting a Python data structure into a JSON string that can be returned as a response from an API endpoint. JSON is a lightweight data-interchange format that is easy to read and write, making it a popular choice for web APIs.

By jsonifying response data, the data can be easily transmitted over HTTP and used by other systems and programming languages. This is because JSON is a language-independent data format that can be parsed and generated by many programming languages.

JSON also supports complex data structures such as arrays and objects, making it a flexible format for transferring data between systems. In a Flask application, the *jsonify* function can be used to convert Python data structures into JSON format.

This function takes the data as an argument and returns a Flask response object with the JSON data and the appropriate `Content-Type` header, indicating that the data is in JSON format. By returning JSON-formatted responses from API endpoints, clients can easily consume and use the data in their applications.

You can see that `content-type` in the preceding code is `text/html`; now, let's return a serialized JSON format since moving forward, that will be the preferred data exchange format:

```
from flask import Flask,jsonify
app = Flask(__name__)
@app.route("/")
def index():
    return "Welcome to Bizza REST API Server"

@app.route("/api/v1/venues")
def venues():
    return jsonify({"id":1,"name":"Auditorium A"}), 404

if __name__ == "__main__":
    app.run()
```

In the preceding code, we added another endpoint and decorated it with `@route("/api/v1/venues")`. So, we are telling Flask to implement the functionality of the view function attached to the decorator.

In order to retrieve the JSON format response, we invoke `jsonify()` in the `Flask` package using `from flask import Flask, jsonify` and pass Python dictionary data into it, which then gets converted into a serializable JSON format.

We can test the endpoint with `curl` by entering the following command:

```
curl.exe http://localhost:5000/api/v1/venues
```

We get the following output:

Figure 9.2 – Screenshot showing the venues endpoint being tested

Next, we will begin to enhance the endpoint functionality by incorporating query parameters.

Adding query parameters to endpoints

Query parameters are additional information we can pass along with the request to the server to allow some processing of the request. With query parameters, we are able to present dynamic content to application users.

For instance, this is an ordinary URL endpoint without a query parameter:

```
http://localhost:5000/
```

Now, you can add a query parameter to the URL by adding a ? at the end of the URL followed by a key-value pair. Let's add a query parameter to the preceding URL:

```
http://localhost:5000/?firstname=Jim&lastname=Hunt
```

Let's implement a simple query parameter in a Flask application for a better illustration.

Add the following snippet to the app.py file:

```python
Import flask, jsonify, request

@app.route("/api/v1/speakers/")
def speakers():
    firstname = request.args.get("firstname")
    lastname = request.args.get("lastname")

    if firstname is not None and lastname is not None:
        return jsonify(message="The speaker's fullname :" +
            firstname+" "+lastname)
    else:
        return jsonify(message="No query parameters in the
            url")
```

Note the following in the preceding code:

- We are creating a new endpoint with the URL /api/v1/speakers/ resource.
- We are using a request object from the Flask package.
- We are then defining a view function, speakers(), to handle requests to the endpoint.
- The request object is used to allow a client to send data to the server and other endpoint request operations. We have request.args for handling URL data, request.form for extracting form data, and request.json for handling JSON data. Here, we will use request.args to extract key-value pairs in the URL to process URL data at the server end.
- The firstname and lastname variables store the data values extracted from the URL.

- Then, we perform a simple check to find out whether query parameters are present in the URL. In the production code, you are expected to perform a comprehensive check on what the users are sending to the server. This is for demonstration purposes. We return JSONified data if query parameters are present. Otherwise, the message `"No query parameters in the url"` is sent as output.

Testing the endpoint with query parameters, for instance, `http://localhost:5000/api/v1/speakers?firstname=Jim&lastname=Hunt`, provides the following output:

```
←  →  C     ⓘ localhost:5000/api/v1/speakers?firstname=Jim&lastname=Hunt

▼ {
     "message": "The speaker's fullname :Jim Hunt"
  }
```

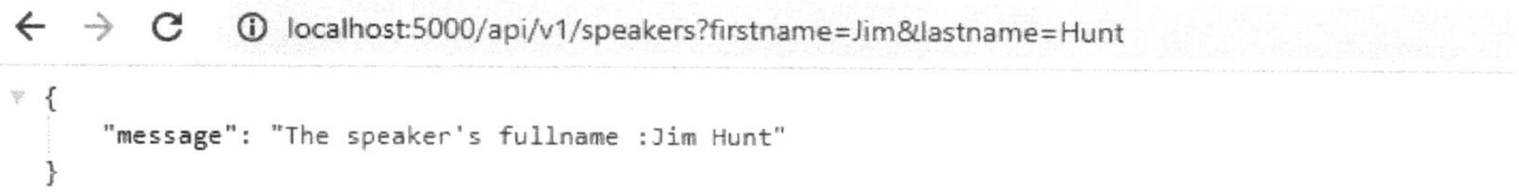

Figure 9.3 – Screenshot testing with a query parameter

Testing the `http://localhost:5000/api/v1/speakers?` endpoint without query parameters provides the following output:

```
←  →  C     ⓘ localhost:5000/api/v1/speakers?

▼ {
     "message": "No query parameters in the url"
  }
```

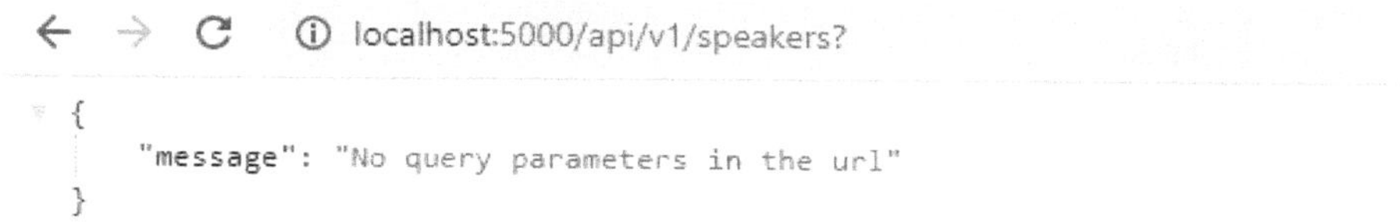

Figure 9.4 – Screenshot testing without a query parameter

Now, let's check how you can pass variables to an endpoint. This is also useful in dynamically modifying the server-side processing in your application.

Passing variables to an endpoint

One of the ways to provide input to an endpoint is by passing variables to the URL path in the API endpoint, allowing for the provision of specific information. This technique is commonly used when the input is required to identify a specific resource or object, such as an ID number or username.

In a Flask application, variables can be included in the URL path by enclosing them in angled brackets (`<>`) in the endpoint URL definition. For example, to define an endpoint that takes a speaker ID as input, the URL could be defined as follows:

```
from flask import Flask, jsonify

app = Flask(__name__)
```

```
@app.route('/api/v1/speakers/<int:speaker_id>')
def get_speaker(speaker_id):
    # Use the speaker ID to fetch the appropriate speaker
      data
    # ...

    # Return the speaker data as a JSON response
    return jsonify(speaker_data)

if __name__ == '__main__':
    app.run()
```

In the preceding example, the get_speaker function takes a speaker_id argument, which corresponds to the variable included in the URL path. When a request is made to the /speakers/123 endpoint, the get_speaker function is called with speaker_id=123.

Passing variables to an endpoint is a useful technique for providing input to an API endpoint and is commonly used in RESTful API design.

Next, we will take our REST API in Flask application a step further. We will perform CRUD operations on a database using PostgreSQL.

API interaction with a database via CRUD operations

In most web application projects, it is common to work with databases for the purpose of persistent data storage. You won't be hardcoding plain text into your REST API, unless you are that person who tries to boil the ocean.

CRUD, an acronym for **create, read, update, and delete**, allows you to manage the state of resources in the database. Interestingly, each of the CRUD elements can also be mapped to the HTTP methods—GET, POST, PUT/PATCH, and DELETE – which further describes and facilitates the interaction with the database.

In a full stack web application, you expect your users to able to create a resource (POST or PUT if it is an existing resource), read a resource (GET), update a resource (PUT/PATCH), and delete a resource (DELETE). In this section, we will work with a simple venue resource with the following endpoints, and the HTTP operations we will perform on them are CRUD.

All the code for the endpoint's implementation will be hosted in the GitHub repository of this book.

Let's jump-start the CRUD operation by describing the endpoints we are using:

- POST /api/v1/venues/: Creates a venue resource
- GET /api/v1/venues: Returns a list of all the venues

- GET `/api/v1/venues/id`: Retrieves a single venue identified with `id`

- PUT `/api/v1/venues/id`: Updates a single venue identified with `id`

- DELETE `/api/v1/venues/id`: Deletes a single venue identified with `id`

The preceding endpoints are clear with regard to their intentions. But before we start fleshing out the endpoints, let's discuss the necessary dependencies and be sure we can connect with the database.

Activate the virtual environment: always remember you are working in a virtual environment to contain your project dependencies. Then, inside `bizza/backend`, update the `app.py` file with the following code:

```
from flask import Flask
from flask_sqlalchemy import SQLAlchemy
app = Flask(__name__)
app.config['SQLALCHEMY_DATABASE_URI'] =
'postgresql://<username>:<password>@localhost:5432/<database_name>'
app.config['SQLALCHEMY_TRACK_MODIFICATIONS'] = False
db = SQLAlchemy(app)
```

In the preceding code snippet, we imported the `Flask` and `SQLAlchemy` modules. Then, the `app = Flask(__name__)` line created an instance of the Flask application. The `__name__` argument represents the name of the current module.

The `Flask` API comes with some configuration settings that we can modify. The `config` object is in the form of a Python dictionary. We are able to set the database URI with `app.config['SQLALCHEMY_DATABASE_URI']` and disable the SQLAlchemy operation notification setting with `app.config['SQLALCHEMY_TRACK_MODIFICATIONS'] = False`.

> **Note**
> In `app.config['SQLALCHEMY_DATABASE_URI'] = 'postgresql://<username>:<password>@localhost:5432/<database_name>'`, change `<username>` and `<password>` to your appropriate database credentials.

With `db = SQLAlchemy(app)`, we created the `SQLAlchemy` instance, which accepts the `Flask` instance as an argument.

With this set, let's define the model class and create the venue table in the database.

Create the `Venue` class model, as follows:

```
class Venue(db.Model):
    __tablename__ = 'venues'
    id = db.Column(db.Integer, primary_key=True)
```

```
name = db.Column(db.String(100))

    def format(self):
        return {
            'id': self.id,
            'name': self.name
        }
```

Open the command terminal and enter the following:

`flask shell`

Then, enter the following:

`from app import db, Venue`

The preceding command brings db, an SQLAlchmy instance, and the Venue class model into scope.

Now, enter the following:

```
db.create_all()
```

The preceding command creates the venues table from the Venue class model. Alternatively, you can run a migration command as done previously to create the table.

Finally, verify the table creation with the following:

```
db.engine.table_names()
```

The following screenshot shows the terminal commands to show the creation of the venues table.

```
C:\Windows\System32\cmd.exe - flask shell
(venv) C:\bizza\backend>flask shell
Python 3.10.1 (tags/v3.10.1:2cd268a, Dec  6 2021, 19:10:37) [MSC v.1929 64 bit (AMD64)] on win32
App: app
Instance: C:\bizza\backend\instance
>>> from app import db, Venue
>>> db.create_all()
>>> db.engine.table_names()
['user', 'venues', 'users']
>>>
```

Figure 9.5 – Screenshot showing commands in flask shell

Now that we have the database and venues table up and running, let's start to define the endpoints.

Creating a venue resource

Let's define the /venues endpoint in app.py and use it to post entries to the database.

The endpoint for creating a new venue is as follows:

```
from flask import Flask, request, jsonify


@app.route("/api/v1/venues", methods=['POST'])
def add_venues():
    if request.method == 'POST':
        name = request.get_json().get('name')
        all_venues =
            Venue.query.filter_by(name=name).first()
        if all_venues:
            return jsonify(message="Venue name already
                exist!"), 409
        else:
            venue = Venue(name=name)
            db.session.add(venue)
            db.session.commit()
            return jsonify({
                'success': True,
                'venues': venue.format()
            }), 201
```

In the preceding code, we have the following:

- The jsonify and request methods are imported from Flask.

- The @app.route() decorator has the URL to the '/api/v1/venues' endpoint with the HTTP POST method.

- The add_venue() function is invoked once there is a POST request.

- We test to make sure request.method == 'POST'.

- We test to ensure the new venue name is not already in the database. If the added name is already in the database, the message "Venue name already exist" is sent back as a json message with status code 409- content conflict.

- If the preceding step fails, we proceed to add to the database session with db.session.add(new_venue). At this point, the entry is not fully added to the database but added to the database session. We need one more step to commit it to the database, which is db.session.commit().

- Jsonify() is an in-built Flask function that returns a JSON-serialized response object.

Returning lists of venues

The following is the endpoint to retrieve all the venues in the database:

```
# retrieve all venues endpoint
@app.route("/api/v1/venues", methods=['GET'])
def retrieve_venues():
    if request.method == 'GET':
        all_venues = Venue.query.all()
        if all_venues:
            return jsonify({
                'success': True,
                'venues': [venue.format() for venue in
                    all_venues]
            }), 200
        return jsonify(message="No venue record found"),
            404
```

In the preceding code, we have the following:

- The GET request method is invoked

- Venue.query.all() is a query from SQLAlchemy to retrieve all the venues

- The jsonify response object output results in a 200 status code, and if it fails, a "No venue record found" message is displayed with status code 404

Returning a single venue resource

The following is an endpoint to return a single venue:

```
@app.route("/api/v1/venues/<int:id>", methods=['GET'])
def retrieve_venue(id):
    if request.method == 'GET':
        venue = Venue.query.filter(Venue.id == id).first()
        if venue:
            return jsonify({
                'success': True,
                'venue': venue.format()
            }), 200
        return jsonify(message="Record id not found"), 404
```

In the preceding code, we are doing the following:

- The GET request method is invoked

- `Venue.query.filter(Venue.id == id).first()` retrieves the first record with the ID supplied as an argument to `retrieve_venue()`

- If the ID is present, the `jsonify` response object output results in a `200` status code, and if it fails, the `"Record id not found"` message is displayed with status code `404`

Updating a single venue resource

The following is an endpoint to modify venue information:

```
@app.route("/api/v1/venues/<int:id>", methods=['PUT'])
def update_venue(id):
    if request.method == 'PUT':
        name = request.get_json().get('name')
        venue = Venue.query.get(id)
        if not venue:
            return jsonify(message='Venue record not
                found'), 404
        venue.name = name
        db.session.commit()
    return jsonify({
        'success': True,
        'updated venue': venue.format()
    }), 200
```

In the preceding code, we are doing the following:

- The PUT request method is invoked to update a single resource

- We try to check for the existence of the record `id`

- If the record is present, we update it with `venue.name = name`, then commit the session

- If the ID information is updated, the `jsonify` response object output results in a `200` status code, and if it fails, the `"Venue record not found"` message is displayed with status code `404`

Deleting a single venue resource

The following is an endpoint to delete a venue:

```
@app.route('/venues/<int:id>', methods=['DELETE'])
def remove_venue(id):
```

```
venue = Venue.query.filter_by(id=id).first()
if venue:
    db.session.delete(venue)
    db.session.commit()
    return jsonify(
        {'success': True,
        'message': 'You deleted a venue',
        'deleted': venue.format()
        }
    ), 202
else:
    return jsonify(message="That venue does not
        exist"), 404
```

In the preceding code, we are doing the following:

- The DELETE request method is invoked to delete a single resource
- We try to check for the existence of the record id
- If the record is present, we delete it, then commit the session
- If the ID information is deleted, the jsonify response object output results in a 202 status code, and if it fails, a "That venue does not exist" message is displayed with status code 404

The following GitHub link contains the complete code for the venue CRUD operation in app. py -https://github.com/PacktPublishing/Full-Stack-Flask-and-React/ blob/main/Chapter09/05/bizza/backend/app.py.

Next, we will examine the concept of API documentation and deep-dive into its importance and how to make use of the Postman tool to document your API endpoint.

API documentation

Have you ever tried to assemble a piece of equipment you bought from a store without looking through the manual? There is a high probability you will have done so deliberately. You smartly think it will be easy to put together, and most times, you get burned in this process of trial and error and eventually mess things up.

Manuals are a good resource meant to ease you through a product's features and functionality. API documentation is no different from manuals for guidance, a set of instructions, references, or even tutorial materials that can enhance developers' understanding of your API.

Why is API documentation a critical component of your API? We will examine a few reasons and dive into using the Postman client tool to document your API so that other developers can understand what it is all about and how to use it. We will use a very simplistic case study with the venues CRUD operation we examined in the preceding section.

In a production-ready API product, you are expected to provide more detailed information about your API in your documentation. You are to write up a technical instructional guide that will make consuming your API services a walk in the park. You would also want to include tutorials, clear references, and sample code that can help developers with easy integration.

Let's examine the reasons why you would want to come up with clear documentation for your API:

- **Enhanced developer experience**: API documentation opens up a way to provide quick-start guides, references, and case studies for developers who want to be productive and consume third-party data to improve their software product offerings. Therefore, well-thought-out API documentation helps developers to understand the best approach to implementing your API data points to solve their problems.

    High-quality and useful API documentation means developers are going to easily understand how to implement your API and thus will increase their overall emotional attachment to your API product. Stripe, PayPal, Spotify, Twilio, and Paystack are good examples of popular commercial APIs with awesome documentation.

- **Decreased onboarding time**: Comprehensive documentation allows even an entry-level developer to quickly know how to implement your API endpoints and methods and handle requests and responses seamlessly without assistance from other developers. This will save time and costs for the business and improve the API adoption rate.

- **Adherence to API objectives**: Documentation allows API designers and consumers to have a common ground on API specs and the problem it aims to solve without ambiguity.

Next, we will use Postman to test and document a RESTful API.

Testing and documenting with Postman

Postman is an advanced API platform that provides a range of features to simplify the development, testing, and utilization of APIs. Postman offers a user-friendly interface for testing API endpoints, generating API documentation, and collaborating on API-related operations.

To begin using Postman, you can download and install the application on your local machine. Visit the official Postman website at https://www.postman.com/downloads to access the download page. From there, you can choose the appropriate version for your operating system and follow the installation instructions.

Once you have Postman installed, you can leverage the power of Postman to document RESTful API endpoints, execute requests, analyze responses, and perform comprehensive testing.

Let's leverage `venues` on the collections and generate API documentation that tells other developers how to make requests and describe the response to expect:

1. On your computer, start Postman and create a collection. Click on the **Collections** tab on the left sidebar of the Postman interface; if it's not visible, you can click on the three horizontal lines in the top-left corner to expand the sidebar. In the dialog box, enter a name for your collection.

Figure 9.6 – Screenshot showing the creation of a collection in Postman

2. After creating a collection, create a new request. Within the collection, click on the **New** button located in the top-left corner of the interface. A drop-down menu will appear:

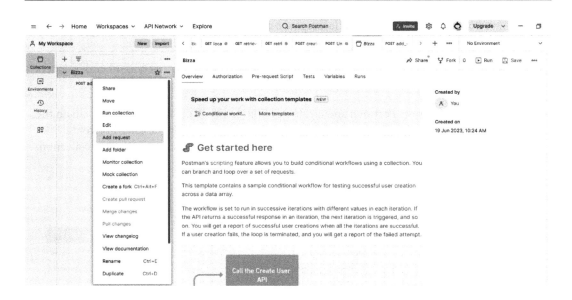

Figure 9.7 – Screenshot showing how to add an HTTP request

3. Select the request type from the drop-down menu. To post data to the server, select **POST** as the request type. This indicates that you want to post data to the server. Then, enter the endpoint: `localhost:5000/api/v1/venues`.

4. Select the **Body** tab, then **raw**, and finally select the **JSON** tab from the dropdown. This will allow you to send the data to the server in JSON format. In this case, we send the following data to the backend server:

    ```
    {
    ''name": "Auditorium A"
    }
    ```

The following screenshot shows how you can use Postman to test endpoints.

Figure 9.8 – Screenshot showing sending data to the backend

5. Send the request once you have configured the request. Click on the **Send** button located on the right side of the request panel. Postman will send the request to the specified endpoint. Make sure your Flask application is running, or else the POST request will fail.

    After sending the request, Postman will display the response received from the server. You can see the response content, status code, headers, and other relevant information in the response panel. The following screenshot shows the response data.

Figure 9.9 – Screenshot showing server response to a POST request

Repeat the preceding steps for each request in your collection, choosing appropriate request types and providing descriptions, parameter details, the request payload, and the response format as needed.

6. Once you have tested all the endpoints, right-click on the collection name and click on **View Documentation**.

This takes you to where you can customize your API documentation further. You can add details such as the API description, stating a brief overview of the API, including its purpose, functionality, and any relevant background information.

Also, you can specify the base URL of the API, including the protocol (HTTP/HTTPS) and the domain name. If the API requires authentication, explain the authentication mechanism(s) supported (for example, API key, OAuth, or JWT) and provide instructions on how to authenticate requests.

Depending on the requirements of your API, you can document each endpoint exposed by the API. For each endpoint, you may include the following information:

- **Endpoint URL**: Provides the URL pattern for the endpoint, including any required path parameters.

- **HTTP method**: Indicates the HTTP method used by the endpoint (`GET`, `POST`, `PUT`, `DELETE`, and so on).

- **Request parameters**: Specifies any query parameters, request headers, or request body parameters required by the endpoint. Include the parameter name, type, description, and whether it is required or optional.

- **Response format**: Describes the format of the response returned by the endpoint (for example, JSON or XML).

- **Response codes**: Lists the possible HTTP status codes returned by the endpoint and their meanings (for example, `200 OK`, `400 Bad Request`, or `401 Unauthorized`).

- **Response body**: Provides an example of the response body returned by the endpoint, including all relevant fields and their descriptions.

- **Error handling**: Explains how errors are handled by the API and provides examples of error responses.

7. Lastly, you can now publish your API documentation to share with the public.

Figure 9.10 – Screenshot showing how you publish your API documentation

You can read up on the further customization of the published documentation in the Postman documentation as it is possible to convert it into HTML and host it on your own server. The link to the generated API documentation is `https://documenter.getpostman.com/view/4242057/2s93sjUoYX`.

Summary

In this chapter, we explored API development in Flask applications. We started by understanding what an API is all about and why businesses and developers are creating and consuming APIs to drive their data access. We took things further by taking a quick run through common terminologies developers come across when implementing the API design pattern.

Then, we unmasked the structure of endpoints and payloads, recognizing that designing routes and exchanging data form the foundational elements of API development. In addition, we critically examined the design principles guiding RESTful API development. We discussed how understanding the REST API design principles enhances best practices in API design and development. Also, we discussed the implementation of the REST API and how we can connect API backend services with a database in a Flask application.

Finally, we discussed API testing and documentation with Postman. In API design and development, we recognized how testing and documenting endpoints is crucial to building stable and usable web applications.

In the next chapter, we are going to bridge the frontend and backend functionalities and experience the full stack nature of React and Flask.

10

Integrating the React Frontend with the Flask Backend

This chapter represents a critical point in our quest to build a full stack web application. In this chapter, you will be introduced to a set of instructions on how to connect a Flask web server to a React frontend. You will learn how to pass form entries from the React frontend to the Flask backend. And after this integration, you can be officially called a **full stack web developer**.

React web applications usually have a sleek look and feel and are regarded as the Rolls-Royce of modern frontend web application applications. React has an intuitive user-interface-focused library, capable of powering production-grade web and mobile applications with ease.

The robust React ecosystem coupled with React's tools and libraries facilitates end-to-end web development. When you combine React's incredible component-based design pattern with a minimalist lightweight Flask framework, you get a rich web application that can withstand the test of time and scale at large.

This chapter will help you understand the dynamics of integrating React, a frontend library, and Flask, a backend framework, in developing valuable software products. You will also learn how React handles forms with the Flask backend in this chapter.

Ever since the advent of the web, there has been a need for more dynamic and responsive forms in web applications. We will explore server-side handling of form elements, validation, and security concerns.

In this chapter, we'll cover the following topics:

- The *Bizza* application structure
- Configuring the React frontend
- Making the Flask backend ready
- Handling forms in React and Flask
- Troubleshooting tips for the React frontend and the Flask backend

Technical requirements

The complete code for this chapter is available on GitHub at: `https://github.com/PacktPublishing/Full-Stack-Flask-and-React/tree/main/Chapter10`.

The Bizza application structure

In this section, we will go deeper into the structure of the application we will build in this book. As stated previously, we will name the fictitious web app *Bizza*, a conference event web application.

This *Bizza* web app will serve as the digital hub for a conference event for speakers from the information technology industry, providing a myriad of features and functionalities that enhance the speakers' and attendees' overall experience. Let's delve into the *Bizza* application structure.

Application overview

Bizza is a fictitious data-driven event application that allows subject experts in the information technology industry to share their insights and experiences, providing valuable knowledge to enhance event attendees' skills.

Bizza lets you see a list of speakers and workshop schedules with details. This site lets users register and browse for a workshop. Essentially, the application will have the following functionalities:

- A home page that displays event speakers and available event schedules with locations and subjects
- A registration form for event attendees
- A registration form for speakers with a subject of interest
- A page for users to log in to the application
- A page containing the names and details of speakers

Next, we will delve into the *Bizza* app and break it down into its frontend and backend components. By doing so, we will gain a comprehensive understanding of the distinct roles and functionalities that each component serves within the app.

Breaking down the code structure into frontend and backend

In the world of software development, frontend and backend are like the yin and yang – opposites but complementing each other to provide a harmonious digital experience. *Yin and yang* is a Chinese philosophical concept that describes opposite but interconnected forces.

In a nutshell, breaking down an application into its frontend and backend components provides a clear separation of concerns, promotes code reusability and portability, enables scalability and performance optimization, and fosters collaboration and parallel development. This approach ultimately contributes to the overall success of the web application development process.

The practice of separating frontend and backend components in software development started to gain prominence in the late 1990s and early 2000s, with the rise of web-based applications. During this time, web technologies were evolving in the blink of an eye, and the need for scalable and modular applications became apparent.

The introduction of JavaScript frameworks such as jQuery in the early 2000s enabled more dynamic and interactive user interfaces in the frontend. This led to a clearer distinction between the presentation layer (frontend) and the data processing layer (backend) of web applications.

With the emergence of **single-page applications** (**SPAs**) and the proliferation of JavaScript frameworks and libraries such as AngularJS, React, and Vue.js, the separation between the frontend and the backend became more standardized and widely adopted. SPAs shifted the responsibility of rendering and managing the UI to the client side, while the backend APIs handled data retrieval and manipulation.

Now that we have discussed the key reasons for breaking down the code structure, let's examine the frontend and backend components of the *Bizza* web application.

The following code structure represents the high-end level code split between the frontend and the backend. This allows us to separate concerns and improve code reusability:

```
bizza/
├── backend/
├── frontend/
```

Frontend structure

First, let's provide an overview of the detailed frontend structure:

```
frontend/

├── node_modules
├── package.json
├── public
    ├──favicon.ico
├──index.html
├── src
    ├──components/
    ├──pages/
    ├──hooks/
    ├──assets/
    └──App.js
    └──App.css
    └──index.js
    └──index.css
    └──setupTests.js
```

```
├──.gitignore
├──.prettierrc
├──package-lock.json
├──package.json
├──README.md
```

The preceding frontend code structure comprises majorly of node_modules, package.json, public, src, .gitignore, .prettierrc, package-lock.json, and README.md.

Let's quickly break down the major directories and files:

- node_modules: This directory contains all the packages (libraries and frameworks) that your application depends on. These packages are listed in the dependencies and devDependencies sections of the package.json file.

- package.json: This file contains metadata about your application, including its name, version, and dependencies. It also includes scripts you can use to build, test, and run your application.

- public: This directory contains static assets that your application will use, such as the favicon and the main HTML file (index.html).

- src: This directory contains the source code for your application. It is organized into subdirectories for components, pages, hooks, and assets. The src directories are critical to the design pattern adopted for the React frontend. The components folders contain all the components we intend to use in the *Bizza* app, pages contains the presentational component of the app, hook contains custom hooks, and finally, the assets folder contains all the assets, such as images, logos, and svg, that are used in the application.

- .gitignore: This file tells Git which files and directories to ignore when you commit your code to a repository.

- .prettierrc: This file specifies the configuration options for the Prettier code formatter. Prettier is a popular code formatting tool that enforces a consistent style across your code base. It is typically placed in the root directory of a JavaScript project and contains JSON syntax to define the formatting rules.

- package-lock.json: This file records the exact versions of all the packages that your application depends on, as well as any packages those packages depend on. It ensures that your application uses the same versions of its dependencies every time it is installed.

- README.md: This file contains documentation for your application, such as instructions to install and run it.

Backend structure

Next, we will examine how the backend will be structured:

```
backend/
├── app.py
├── models
├── config
│    ├── config.py
├── .flaskenv
├── requirements.txt
```

The preceding represents the files and directories structure for the Flask backend application.

Let's break down the directories and files:

- `app.py`: This file contains the main code for your backend application, including routes and logic to handle HTTP requests.

- `models`: This directory contains modules of each of the model definitions for database models.

- `config`: This directory contains a configuration options file for the application, such as database connection strings or secret keys.

- `.flaskenv`: This file contains environment variables that are specific to the Flask application.

- `requirements.txt`: This file lists the packages that the application depends on, including any third-party libraries. You can use this file to install the necessary dependencies by running `pip install -r requirements.txt`.

Next, we will see how to configure the React frontend and prepare it to consume the backend API services.

Configuring the React frontend for API consumption

In this section, you will configure the frontend React app to communicate with the backend Flask server by setting up a proxy in React to consume the API from the Flask server.

In order to configure the React proxy for API consumption, you will need to update the `proxy` field in the `package.json` file of the frontend React app. The `proxy` field allows you to specify a URL that will be used as the base for all API requests made from the React app.

Let's update the `package.json` file:

1. Open the `package.json` file in the `project` directory using a text editor, and then add a `proxy` field to the `package.json` file and set it to the URL of your Flask server:

    ```
    {
      "name": "bizza",
    ```

```
    "version": "0.1.0",
    "proxy": "http://localhost:5000"
}
```

2. Next, you will need to make HTTP requests to the Flask server from the React frontend. We will use the **Axios** library for this purpose. The `Fetch()` method is an alternative to Axios.

    Axios is a JavaScript library that allows you to make HTTP requests from the browser. It is a promise-based library that uses modern techniques to make it easy to work with asynchronous requests. With Axios, you can make HTTP requests to retrieve data from a server, submit form data, or send data to a server.

    Axios supports a number of different request methods, such as `GET`, `POST`, `PUT`, `DELETE`, and `PATCH`, and it can handle both JSON and XML data formats. Axios is popular among developers because it has a simple and straightforward API, making it easy to use for both beginners and experienced developers.

    Axios also has a number of features that make it flexible and powerful, such as the automatic transformation of data, support for interceptors (which allow you to modify requests or responses before they are sent or received), and the ability to cancel requests.

3. You can install Axios by running the following command in your terminal:

    ```
    npm install axios
    ```

    Once Axios is installed, you can use it to make HTTP requests to the Flask server from the React frontend.

4. Make sure that both the frontend React app and the backend Flask server run on separate ports. By default, the React development server runs on port `3000`, while the Flask development server runs on port `5000`.

Next, you will need to define routes and functions in the Flask backend to handle HTTP requests coming from the React frontend.

Making Flask backend-ready

In the *Setting up the development environment with Flask* section of *Chapter 1*, *Getting Full Stack Ready with React and Flask*, we set up the development environment for the Flask server. Ensure your virtual environment is activated. You can do so by running the following commands:

*   **For Mac/Linux**:

    ```
    source venv/bin/activate
    ```

*   **For Windows**:

    ```
    Venv/Scripts/activate
    ```

Your virtual environment should now be activated, and your terminal prompt should be prefixed with the name of the virtual environment (for example, (venv) $).

Next, let's dive straight into defining an event registration route, with its function as part of the requirements for Bizza application model.

Let's add a model to handle registrations for event attendees. You will later use it to accept requests from the React frontend in the next section, where we will handle form inputs in React and Flask.

The app.py file in the root directory of your application is still the main entry point for the Flask application. Update app.py with the following code snippet to define the model and endpoint to handle the event registration:

```python
class EventRegistration(db.Model):
    __tablename__ = 'attendees'
    id = db.Column(db.Integer, primary_key=True)
    first_name = db.Column(db.String(100), unique=True, nullable=False)
    last_name = db.Column(db.String(100), unique=True, nullable=False)
    email = db.Column(db.String(100), unique=True, nullable=False)
    phone = db.Column(db.String(100), unique=True, nullable=False)
    job_title = db.Column(db.String(100), unique=True, nullable=False)
    company_name = db.Column(db.String(100), unique=True,
        nullable=False)
    company_size = db.Column(db.String(50), unique=True,
        nullable=False)
    subject = db.Column(db.String(250), nullable=False)


def format(self):

    return {
        'id': self.id,
        'first_name': self.first_name,
        'last_name': self.last_name,
        'email': self.email,
        'phone': self.phone,
        'job_title': self.job_title,
        'company_name': self.job_title,
        'company_size': self.company_size,
        'subject': self.subject

    }
```

In the preceding snippet, the EventRegistration class represents a model for event registration in a database.

The `__tablename__` attribute specifies the name of the table in the database that this model is stored in. The `db.Model` class is a base class for all models in `Flask-SQLAlchemy`, and the `db.Column` objects define the fields of the model, each with a type and some additional options.

The `format` method returns a dictionary representation of the model instance, with keys corresponding to the field names and values corresponding to the field values.

Now, let's define the route or endpoint, `/api/v1/events-registration`:

```python
@app.route("/api/v1/events-registration", methods=['POST'])
def add_attendees():
    if request.method == 'POST':
        first_name = request.get_json().get('first_name')
        last_name = request.get_json().get('last_name')
        email = request.get_json().get('email')
        phone = request.get_json().get('phone')
        job_title = request.get_json().get('job_title')
        company_name = request.get_json().get('company_name')
        company_size = request.get_json().get('company_size')
        subject = request.get_json().get('subject')

        if first_name and last_name and email and phone and subject:
            all_attendees = EventRegistration.query.filter_by(
                email=email).first()
            if all_attendees:
                return jsonify(message="Email address already
                    exists!"), 409
            else:
                new_attendee = EventRegistration(
                    first_name = first_name,
                    last_name = last_name,
                    email = email,
                    phone = phone,
                    job_title = job_title,
                    company_name = company_name,
                    company_size = company_size,
                    subject = subject
                )
                db.session.add(new_attendee)
                db.session.commit()
                return jsonify({
                    'success': True,
                    'new_attendee': new_attendee.format()
                }), 201
        else:
            return jsonify({'error': 'Invalid input'}), 400
```

The /api/v1/events-registration endpoint function handles an HTTP POST request to the /api/v1/events-registration route. This endpoint allows users to register for events by providing their name, email address, phone number, and subject.

The endpoint function first checks that the request method is indeed POST, and then extracts the name, email, phone, and subject values from the request body, which is expected to be in JSON format.

Next, the function checks that all of the required input values (first_name, last_name, email, phone, and subject) are present. If they are, it checks whether there is already an attendee with the same email address in the database. If there is, it returns a JSON response, with a message indicating that the email address is already in use, and an HTTP 409 status code (conflict).

If the email address is not in use, the function creates a new EventRegistration object with the input values, adds it to the database session, and commits the changes to the database. It then returns a JSON response with a success message and the details of the new attendee, along with an HTTP 201 status code (created).

If any of the required input values is missing, the function returns a JSON response with an error message and an HTTP 400 status code (bad request). Now, let's update the database and add an eventregistration table to it. The eventregistration table will accept all the entries for the event registrations.

The following steps create the eventregistration table in the database. In the terminal of the project directory, enter the following commands:

```
flask shell
from app import db, EventRegistration
db.create_all()
```

Alternatively, you can continue to use migration tools:

```
flask db migrate -m "events attendee table added"
flask db upgrade
```

With any of these options, the backend will contain the new table.

Execute flask run in the terminal to start the Flask development server on localhost using the default port (5000).

That's it! The backend is now ready to receive form entries from the React frontend. Let's design the form component in React and submit the form entries to the Flask backend.

Handling forms in React and Flask

Handling forms in a React frontend and a Flask backend is a common pattern in web development. In this pattern, the React frontend sends HTTP requests to the Flask backend to submit or retrieve form data.

On the React frontend side, you can use a form component to render a form and handle form submissions. You can use controlled components, such as `input`, `textarea`, and `select`, to control the form values and update the component state as a user inputs data.

When the user submits the form, you can use an event handler to prevent the default form submission behavior and send an HTTP request to the Flask backend using a library such as Axios. In this section, we will work with the Axios library.

On the Flask backend side, you can define a route to handle the HTTP request and retrieve the form data from the request object. You can then process the form data and return a response to the frontend.

The `EventRegistration` component provides a simple form for non-authenticated users to register for an event in the frontend of the *Bizza* app. The form includes fields for the user's name, email address, phone number, and subject – the topic or title of the event they register for.

Let's dive into the React form implementation that will work with the Flask backend:

1. In the project directory and inside the `components` folder, create `EventRegistration/EventRegistration.jsx`.

2. Add the following code snippet to the `EventRegistration.jsx` file:

```jsx
import React, { useState, useEffect } from 'react';
import axios from 'axios';
const EventRegistration = () => {
  // Initial form values
  const initialValues = {
    firstname: '',
    lastname: '',
    email: '',
    phone: '',
    job_title: '',
    company_name: '',
    company_size: '',
    subject: '' };

  // State variables
  const [formValues, setFormValues] =
    useState(initialValues); // Stores the form field
                             values
```

```
    const [formErrors, setFormErrors] = useState({});
// Stores the form field for the validation errors
    const [isSubmitted, setIsSubmitted] =
      useState(false); // Tracks whether the form has
                         been submitted
{/* Rest of the form can be found at the GitHub link - https://
github.com/PacktPublishing/Full-Stack-Flask-Web-Development-
with-React/tree/main/Chapter-10/ */}
               <div id="btn-section">
                  <button>Join Now</button>
               </div>
            </form>
          </div>
        </div>
     </div>
   </>
   );
   };


   export default EventRegistration;
```

Refer to GitHub to get the full source code.

In the preceding snippet, when the form is submitted, the component sends an HTTP POST request to the /api/v1/events-registration route with the form data. It then updates the component's state with the response from the server and displays a success or error message to the user.

The EventRegistration component also includes a validate function that checks the form values for errors, and an onChangeHandler function that updates the form values as the user types.

Let's discuss the component state variables used in the preceding code:

- formValues: This is an object that stores the current values of the form fields (name, email, phone, and subject)

- formErrors: This is an object that stores any errors found in the form values

- response: This is an object that stores the response from the server after the form is submitted

- feedback: This is a string that stores a feedback message to display to the user (for example, **Registration successful!**)

- status: This is a string that stores the status of the form submission (for example, **success** or **error**)

We then define the following functions:

- `validate`: This is a function that accepts the form values and returns an object with any errors found in the values.
- `onChangeHandler`: This is a function that updates the `formValues` state variable as the user types in the form fields.
- `handleSubmit`: This is a function that is called when the form is submitted. It prevents the default form submission behavior, calls the `validate` function to check for errors, and then sends the form data to the server using the `sendEventData` function. It also updates the feedback and status state variables based on the response from the server.
- `sendEventData`: This is an `async` function that sends an HTTP `POST` request to the `/api/v1/events-registration` route with the form data and updates the response state variable with the response from the server.

The `EventRegistration` component also has a `useEffect` hook that calls the `sendEventData` function when the `formValues` state variable changes. Finally, the `EventRegistration` component renders a form element with the form fields and displays the feedback message and status to the user.

Now, start the React frontend with `npm start` and submit your form entries. Make sure the Flask server is running as well. With any development process, issues and bugs are bound to arise. We will explore some valuable troubleshooting tips to help you debug and fix issues during the integration of the React frontend and the Flask backend.

Troubleshooting tips for the React frontend and the Flask backend

Integrating a React frontend with a Flask backend can be a powerful combination to build dynamic and scalable web applications. However, like any integration, it can come with its own set of unavoidable challenges. Troubleshooting issues that arise during the React-Flask integration process requires a systematic approach to identify and resolve problems effectively.

This section will discuss how you can resolve some of the issues that you may encounter while integrating the frontend with the backend. By following these tips, you will be able to diagnose and resolve common problems that may arise during the development and deployment of your application.

Let's dive into the troubleshooting tips for the React frontend and the Flask backend integration:

- **Verifying the Flask setup**:
  - Ensure that Flask is properly configured and running on the server
  - Check the Flask server console for any error messages or exceptions that may indicate misconfigurations

- Confirm that the necessary Flask packages and dependencies are installed

- Verify that the Flask server is accessible and responding to requests by testing basic endpoints

- **Checking the React configuration**:

  - Ensure that the React application is correctly configured and running

  - Confirm that the necessary dependencies and packages are installed in the React project

  - Inspect the console logs in the browser's developer tools for any JavaScript errors or warnings that may indicate issues with the frontend setup

  - Ensure that the proxy attribute is added to `package.json` and points to the Flask server address – for example, `http://127.0.0.1:5000`

- **Investigate network requests**:

  - Use the browser's developer tools to inspect network requests made by the React application

  - Verify that the requests are sent to the correct Flask endpoints

  - Check the network response status codes to identify any server-side errors

  - Inspect the response payloads to ensure that data is transmitted correctly

  - Pay attention to **Cross-Origin Resource Sharing (CORS)** issues if the React frontend and the Flask backend are hosted on different domains or ports

By following these troubleshooting tips, you will be equipped with the necessary knowledge to efficiently diagnose and resolve issues in the React-Flask integration. This will ensure a smooth and robust integration for your web application.

Summary

In this chapter, we discussed extensively the application code structure and a few key steps required to integrate a React frontend with a Flask backend. First, you will need to set up the frontend to communicate with the backend, using an HTTP client library, and handle forms and user input.

Then, you will need to set up the Flask backend with the necessary routes and functions to handle the requests from the frontend and process the form data. Finally, you will need to test the entire application to ensure that it works correctly and as expected.

With these steps, you can successfully integrate a React frontend with a Flask backend for your web application. In the next chapter, we will scale up the React-Flask interactions by creating more tables. These tables will have relationships, and we will be able to fetch and display data.

11

Fetching and Displaying Data in a React-Flask Application

In the preceding chapter, you were able to successfully integrate the React frontend into the Flask backend. This is a significant milestone in the journey of a full stack web developer. In this chapter, you will build on what you have learned and dive deeper into data fetching in a full stack web application.

Data fetching is important in a web application because it allows the application to retrieve data from a backend server, API, or database and display that data to a user. Without the ability to fetch data, a web application would be limited to displaying only hardcoded data, which would not be very useful or dynamic. By fetching data from a backend server or API, the application can display up-to-date, dynamic data to the user.

In addition, data fetching is often used in combination with user interactions and updates to the data, allowing the application to perform actions such as inserting, updating, or deleting data in a database or API. This allows the application to be more interactive and responsive to the user's actions.

In this chapter, you will learn about the intricacies of data fetching and its vital role in web applications, and more importantly, how it concerns integrating the React frontend with the Flask backend. You will learn about the role of data fetching in enabling web applications to fetch data from a backend server or API, ensuring the display of current and dynamic information.

We will discuss the use of data fetching in combination with user interactions to perform actions such as retrieving, inserting, updating, or deleting data in a database or API. Lastly, we will discuss how you can manage pagination in React–Flask applications.

By the end of this chapter, you will understand how to add data to a database, display database data, and how pagination is handled in a React–Flask web application.

In this chapter, we'll be covering the following topics:

- Fetching and displaying data – the React–Flask approach
- Adding data to a database – the React–Flask approach
- Editing data – the React–Flask approach
- Deleting data from a database – the React–Flask approach
- Managing pagination in a React–Flask application

Technical requirements

The The complete code for this chapter is available on GitHub at: `https://github.com/PacktPublishing/Full-Stack-Flask-and-React/tree/main/Chapter11`.

Owing to the page-count constraints, some of the code blocks have been snipped. Please refer to GitHub for the complete code.

Fetching and displaying data – the React-Flask approach

In this chapter, first, we will be fetching data on speakers and displaying it to the users of the application. But before heading into that, let's do some code restructuring. You will need to restructure the backend to accommodate the growing `app.py` file contents in the project directory. Dividing the code into different components improves the overall structure and organization of the application.

Rather than having the entire code in a single module, you can structure your code to separate concerns. We'll discuss more on code structuring for larger applications in *Chapter 14, Modular Architecture – The Power of Blueprints*. With this code split, developers can easily locate and modify specific parts of the code base without affecting other components. This modular approach also promotes code reusability.

Now, back to the code, you will add `models.py` to the backend project directory (`bizza/backend/models.py`) to house all the models for database interaction. This will help us to separate application concerns. The `app.py` file will be used to handle endpoints and their associated logic, while the `models.py` file contains the application data models.

The restructured `app.py` and `models.py` files can be found on GitHub at `https://github.com/PacktPublishing/Full-Stack-Flask-and-React/tree/main/Chapter11`.

Essentially, we will simulate an admin page for our *Bizza* application so that we can create, display, and edit speaker data, and manage pagination via the admin page. At this point, we are setting up an admin page for demonstration purposes only; we are not going to bother ourselves with data validation, authentication, and authorization implementations.

In this section, the focus will be to learn how to retrieve data from the backend and display it in the React frontend. Being able to display data from a database is important because it allows you to present the data to users in a visual and interactive way. By displaying data in a web application, you can create a user-friendly interface that allows users to view, search, filter, and manipulate the data as needed.

You need to fetch and display data in order to create a functional and useful web application that makes use of data stored in a database. To retrieve data from the backend, we will use Axios for making network requests. You can use Axios to make a `GET` request to the backend server and retrieve the data you need.

Let's delve into how we can retrieve a list of speakers and their details from the backend and display it in the admin page of our *Bizza* application.

Retrieving the speakers' list from Flask

The Flask backend will manage the list of speakers and their details with a simple API. In `models.py`, add the following code to create the `Speaker` model class:

```python
from datetime import datetime
class Speaker(db.Model):
    __tablename__ = 'speakers'

    id = db.Column(db.Integer, primary_key=True)
    name = db.Column(db.String(100), nullable=False)
    email = db.Column(db.String(100), nullable=False)
    company = db.Column(db.String(100), nullable=False)
    position = db.Column(db.String(100), nullable=False)
    bio = db.Column(db.String(200), nullable=False)
    speaker_avatar = db.Column(db.String(100),
        nullable=True)
    created_at = db.Column(db.DateTime,
        default=datetime.utcnow)
    updated_at = db.Column(db.DateTime,
        default=datetime.utcnow, onupdate=datetime.utcnow)

    def __repr__(self):
        return f'<Speaker {self.name}>'

    def serialize(self):
        return {
            'id': self.id,
            'name': self.name,
            'email': self.email,
            'company': self.company,
```

```
                      'position': self.position,
                      'bio': self.bio,
                      'speaker_avatar': self.speaker_avatar,
                      'created_at': self.created_at.isoformat(),
                      'updated_at': self.updated_at.isoformat()
            }
```

The preceding code defines a `Speaker` model and has the `__repr__()` and `serialize()` methods. The `__repr__` method is a built-in method in Python that is used to create a string representation of an object. In this case, it is used to create a string representation of a `Speaker` object.

The `serialize()` method is used to convert the `Speaker` object into a dictionary format that can be easily converted into JSON. This is useful when you need to return the `Speaker` object as a response to an API endpoint.

The method returns a dictionary containing all the properties of the `Speaker` object such as `id`, `name`, `email`, `company`, `position`, `bio`, `speaker_avatar`, `created_at`, and `updated_at`. The `created_at` and `updated_at` properties are converted into string format using the `isoformat()` method.

Now, let's create the endpoint to handle the logic for displaying the speakers' data:

```
@app.route('/api/v1/speakers', methods=['GET'])
def get_speakers():
    speakers = Speaker.query.all()
    if not speakers:
        return jsonify({"error": "No speakers found"}), 404
    return jsonify([speaker.serialize() for speaker in
        speakers]), 200
```

The preceding code retrieves the list of speakers from the database with the `get_speakers()` function. Now, you need to update the React frontend directory to consume the API speakers list endpoint.

Displaying data in React

In the React frontend, you need to create a route that renders a component at the `http://127.0.0.1:3000/admin` path.

The following snippet will create the routing system for the admin:

```
const router = createBrowserRouter([
  {
    path: "/admin",
    element: <AdminPage/>,
```

```
  children: [


    {
      path: "/admin/dashboard",
      element: <Dashboard />,
    },
    {
      path: "/admin/speakers",
      element: <Speakers />,
    },
    {
      path: "/admin/venues",
      element: <Venues />,
    },
    {
      path: "/admin/events",
      element: <Events />,
    },
    {
      path: "/admin/schedules",
      element: <Schedules />,
    },
    {
      path: "/admin/sponsors",
      element: <Sponsors />,
    },
    ],



  },
]);
```

Let's now create AdminPage in the /src/pages/Admin/AdminPage/AdminPage.jsx file. AdminPage will serve as an index component page for the admin and render necessary components, including the CRUD operations for speakers.

Add the following code to AdminPage.jsx:

```
import React from "react";
import { Outlet } from "react-router-dom";
import Sidebar from
    "../../../components/admin/Sidebar/Sidebar";
```

```
import './AdminPage.css'

const AdminPage = () => {
    return (
        <div className="container">
            <div><Navbar/></div>
            <div><Outlet /></div>
        </div>
    );
};
export default AdminPage;
```

The preceding code shows the `AdminPage` component, which represents the structure and content of the `admin` page. The `Sidebar` component is imported and rendered as a child component to render a list of sidebar menus for the admin. Then, we have the `Outlet` component imported from the `react-router-dom` package, which is used to render the content specific to the current route.

Next, we will create a data-fetching component for viewing the list of speakers in the database.

Displaying the speakers list with the ViewSpeakers component

We will start the CRUD operations on speakers with the `ViewSpeakers` component that will handle the display of the speakers' data from the backend to the admin users.

First, we will create a module named `SpeakersAPI.js` to handle all the API calls. The `SpeakersAPI.js` module encapsulates the API calls, abstracting away the low-level details of making HTTP requests. This will also allow other parts of the application to interact with the API in a more straightforward manner, without having to deal with the intricacies of the Axios library directly. Overall, you tend to benefit from having this separate module handle the API calls as it promotes code organization, reusability, error handling, header management, and the scalability and maintainability of the code base.

Let's now dig into the `SpeakersAPI` module.

Inside the `bizza/frontend/src` project directory, create `SpeakersAPI.js` and add the following snippet:

```
import axios from 'axios';
const API_URL = 'http://localhost:5000/api/v1';
// Function to handle errors
const handleErrors = (error) => {
    if (error.response) {
    // The request was made and the server responded with a
       status code
    console.error('API Error:', error.response.status,
```

```
        error.response.data);
    } else if (error.request) {
    // The request was made but no response was received
    console.error('API Error: No response received',
        error.request);
    } else {
    // Something else happened while making the request
    console.error('API Error:', error.message);
    }
    throw error;
};
// Function to set headers with Content-Type:
   application/json
const setHeaders = () => {
    axios.defaults.headers.common['Content-Type'] =
        'application/json';
};
// Function to get speakers
export const getSpeakers = async () => {
    try {
        setHeaders();
        const response =
            await axios.get(`${API_URL}/speakers`);
        return response.data;
    } catch (error) {
        handleErrors(error);
    }
};
```

The preceding code sets up a basic configuration for making HTTP requests to an API using Axios and provides a function to retrieve speakers from the API. It handles errors and sets the necessary headers for the requests.

Next, we will define the `ViewSpeakers` component and make use of the preceding `SpeakersAPI` module.

Inside `src/pages/Admin/Speakers/`, create the `ViewSpeakers.js` component and add the following code:

```
import React, { useEffect, useState } from 'react';
import { getSpeakers } from
    '../../../services/SpeakersAPI';
const ViewSpeakers = () => {
    const [speakers, setSpeakers] = useState([]);
```

```
const [isLoading, setIsLoading] = useState(true);
const [error, setError] = useState(null);
const fetchSpeakers = async () => {
    try {
        const speakerData = await getSpeakers();
        setSpeakers(speakerData);
        setIsLoading(false);
    } catch (error) {
        setError(error.message);
        setIsLoading(false);
    }
};
useEffect(() => {
    fetchSpeakers();
}, []);
```

The preceding code sets up a React component called ViewSpeakers that fetches speaker data using the getSpeakers function and updates the component's state accordingly. It handles loading and error states and triggers the data-fetching process when the component is mounted. The full code for ViewSpeakers.js can be found in the GitHub repository.

Next, we will explore how we can add data to the database using the Flask–React approach.

Adding data to a database – the React–Flask approach

We add data to a database to store and organize information that can be easily accessed, managed, and updated. It is one of the ways to persistently store data, and knowing how to do it is a key requirement for any full stack developer. This knowledge allows you to build dynamic and interactive web applications. You then have the means to efficiently retrieve and use the data for various purposes, such as reporting, analysis, and decision making.

Adding data to Flask

Now, let's create an endpoint to handle the logic for adding speaker data to the database:

```
@app.route('/api/v1/speakers', methods=['POST'])
def add_speaker():
    data = request.get_json()
    name = data.get('name')
    email = data.get('email')
    company = data.get('company')
    position = data.get('position')
    bio = data.get('bio')
    avatar = request.files.get('speaker_avatar')
```

```python
    # Save the uploaded avatar
    if avatar and allowed_file(avatar.filename):
        filename = secure_filename(avatar.filename)
        avatar.save(os.path.join(app.config[
            'UPLOAD_FOLDER'], filename))
    else:
        filename = 'default-avatar.jpg'
    if not name or not email or not company or not
        position or not bio:
        return jsonify({"error": "All fields are
            required"}), 400

    existing_speaker =
        Speaker.query.filter_by(email=email).first()
    if existing_speaker:
        return jsonify({"error": "Speaker with that
            email already exists"}), 409
    speaker = Speaker(name=name, email=email,
        company=company, position=position, bio=bio,
            speaker_avatar=avatar)
    db.session.add(speaker)
    db.session.commit()
    return jsonify(speaker.serialize()), 201
# Function to check if the file extension is allowed
def allowed_file(filename):
    return '.' in filename and \
        filename.rsplit('.', 1)[1].lower(
        ) in app.config['ALLOWED_EXTENSIONS']
```

The preceding code defines a /api/v1/speakers route that defines an API route that handles a POST request to add a new speaker. It extracts the required speaker information from the request, validates the data, saves the avatar file if provided, checks for duplicate emails, creates a new speaker object, adds it to the database, and returns a response with the created speaker's data.

The preceding code shows the add_speaker function that is executed when a POST request is made to the specified route.

The add_speaker function retrieves the JSON data from the request using request.get_json() and extracts the speaker's name, email, company, position, bio, and speaker_avatar (an uploaded file) from the data.

If a speaker_avatar is provided and the file extension is allowed (after being checked by the allowed_file function), the avatar file is saved to the server's upload folder with a secure filename. Otherwise, a default avatar filename is assigned.

The function then checks whether all the required fields (name, email, company, position, and bio) have been provided. If any of the fields are missing, it returns a JSON response with an error message and a status code of 400 (bad request).

Next, the add_speaker() function queries the database to check whether a speaker with the same email already exists. If a speaker with the same email is found, it returns a JSON response with an error message and a status code of 409 (Conflict).

If the speaker is new (no existing speaker with the same email), a new Speaker object is created with the provided information, including the avatar file. The speaker is then added to the database session and committed.

Finally, the add_speaker() function returns a JSON response with the serialized speaker data and a status code of 201 (Created) to indicate a successful speaker creation. The code also includes a helper function, allowed_file, that checks whether a given filename has an allowed file extension based on the application's configuration.

Next, we are going to set up the React component to add the speaker data to the backend.

Adding the speaker data to the backend using the CreateSpeaker component

In this section, we will add speaker data to the backend. We will create a component called CreateSpeaker. This component will handle the form inputs for adding a new speaker and send the data to the backend API for storage.

First, we will add the AddSpeaker function to the API call service module SpeakersAPI.js:

```
// API function to add a speaker
export const addSpeaker = (speakerData) => {
    const url = `${API_URL}/speakers`;
    return axios
        .post(url, speakerData, { headers: addHeaders() })
        .then((response) => response.data)
        .catch(handleErrors);
};
```

The preceding code provides an addSpeaker function that utilizes Axios to send a POST request to the backend API for adding a new speaker. It handles the request, response, and error cases appropriately.

Now we will create the CreateSpeaker.js component inside src/pages/Admin/Speakers and add the following code:

```
import React, { useState } from 'react';
import { addSpeaker } from
    '../../../services/SpeakersAPI'LP;
```

```
import { useNavigate } from 'react-router-dom';

const CreateSpeaker = () => {
    const [name, setName] = useState('');
{/* Rest of inputs states */}
    const [isLoading, setIsLoading] = useState(false);
    const [error, setError] = useState(null);
    const [successMessage, setSuccessMessage] =
        useState('');
    const navigate = useNavigate();

    const handleSubmit = async (event) => {
        event.preventDefault();
        setIsLoading(true);
        setError(null);
        try {
            ...

            await addSpeaker(formData);
            setIsLoading(false);

            // Reset the form fields
            setName('');
            setEmail('');
            setCompany('');
            setPosition('');
            setBio('');
            setAvatar(null);

            // Display a success message
            ...
        )}
        </div>
    );
};

export default CreateSpeaker;
```

The preceding code defines a `CreateSpeaker` component that handles the creation of a new speaker. It manages form input values, avatar file selection, loading state, error messages, and success messages. When the form is submitted, the component sends the data to the backend API and handles the response accordingly:

- The component imports necessary dependencies, including `React`, the `useState` hook, the `addSpeaker` function from `SpeakersAPI`, and the `useNavigate` hook from `react-router-dom`.

- Inside the `CreateSpeaker` component, it sets up state variables using the `useState` hook to store the form input values (name, email, company, position, and bio), the avatar file, loading state, error message, and success message. The `CreateSpeaker` component also uses the `useNavigate` hook to handle navigation.

- The component defines a `handleSubmit` function that is triggered when the form is submitted. It first prevents the default form submission behavior. Then, it sets the loading state to true and clears any previous error messages. Within the `handleSubmit` function, the component constructs a `FormData` object and appends the form input values and the avatar file to it.

- The `addSpeaker` function (imported from `SpeakersAPI`) is called with the constructed `FormData` object, which sends a `POST` request to the backend API to create a new speaker.

- If the request is successful, the loading state is set to false, and the form input values are reset. A success message is displayed, and the user is navigated to the `/speakers` page. If an error occurs during the API request, the loading state is set to false, and the error message is stored in the state.

- The component also includes a `handleAvatarChange` function to update the avatar state variable when a file is selected in the avatar input field.

- The component's rendered function returns JSX elements, including a form with form inputs and a submit button. It also displays error and success messages based on the respective state variables.

Now, let's move to the next section to explore how data can be edited in a React–Flask application.

Editing data – the React–Flask approach

In addition to displaying and adding data, it's also important for a web application to allow the user to edit data. In this section, you will learn how to implement data editing in a React-Flask web application.

Editing data in Flask

Now, let's add the endpoint to handle the logic for updating the speaker data in the database. Add the following code to `app.py`:

```python
from flask import jsonify, request
from werkzeug.utils import secure_filename
@app.route('/api/v1/speakers/<int:speaker_id>',
    methods=['PUT'])
def update_speaker(speaker_id):
    data = request.get_json()
    name = data.get('name')
    email = data.get('email')
    company = data.get('company')
    position = data.get('position')
```

```
bio = data.get('bio')
avatar = request.files.get('speaker_avatar')

speaker = Speaker.query.get(speaker_id)

if not speaker:
    return jsonify({"error": "Speaker not found"}), 404

if not all([name, email, company, position, bio]):
    return jsonify({"error": "All fields are
        required"}), 400

if email != speaker.email:
    existing_speaker =
        Speaker.query.filter_by(email=email).first()
```

The preceding code defines a new route for updating a speaker's information at the /api/v1/ speakers/int:speaker_id endpoint that accepts a PUT request. The @app.route decorator is used to define the endpoint, and the methods parameter is set to ['PUT'] to specify that this route should only accept PUT requests. The <int:speaker_id> part of the endpoint is a path parameter, which allows the route to accept a speaker ID as part of the URL.

The code defines the update_speaker function, which takes in a speaker_id parameter that corresponds to the path parameter in the endpoint.

The code first gets the JSON payload of the request and extracts the speaker's information from it. Then, you retrieve the speaker's information from the database using the Speaker.query. get(speaker_id) method. The function queries the database to retrieve the existing speaker object based on the provided speaker_id. If the speaker is not found, it returns a JSON response with an error message and a status code of 404 (Not Found).

update_speaker() checks whether all the required fields (name, email, company, position, and bio) have been provided. If any of the fields are missing, it returns a JSON response with an error message and a status code of 400 (Bad Request).

If there is an exception in saving the image, it will delete the previous avatar image and return an error message and status code. The update_speaker function then updates the speaker's information in the database. The update_speaker function attempts to commit the changes to the database; if it fails, it will roll back the transaction and return an error message and status code of 500.

Finally, if all goes well, the code returns the updated speaker's information as a JSON object and a status code of 200.

Next, we will create the React component to handle updating a speaker's data.

Displaying the edited data in React

In this section, we will provide the functionality to edit speakers' information. To edit data in React, we can modify the component's state with the updated values and reflect those changes in the user interface. We will start by adding the UpdateSpeaker component. In frontend/src/pages/ Admin/Speakers/UpdateSpeaker.js, add the following code:

```
import React, { useState, useEffect } from 'react';
import { updateSpeaker } from
    '../../../services/SpeakersAPI';
import { useNavigate } from 'react-router-dom';
const UpdateSpeaker = ({ speakerId }) => {
    const [name, setName] = useState('');
    const [email, setEmail] = useState('');
    const [company, setCompany] = useState('');
    const [position, setPosition] = useState('');
    const [bio, setBio] = useState('');
    const [isLoading, setIsLoading] = useState(false);
    const [error, setError] = useState(null);
    const [successMessage, setSuccessMessage] =
        useState('');
    const navigate=useNavigate();
    useEffect(() => {
        // Fetch the speaker data based on speakerId
        fetchSpeaker();
    }, [speakerId]);
    const fetchSpeaker = async () => {
        try {
            // Fetch the speaker data from the backend
                based on speakerId

            const speakerData =
                await getSpeaker(speakerId);
            setName(speakerData.name);
            setEmail(speakerData.email);
            setCompany(speakerData.company);
            setPosition(speakerData.position);
            setBio(speakerData.bio);
        } catch (error) {
            setError(error.message);
        }
    };
{/* The rest of the code snippet can be found on GitHub */}
```

The preceding code defines a component called `UpdateSpeaker`. The component allows a user to update the information of a speaker by making a PUT request to the server using the `updateSpeaker` function from the `SpeakersAPI.js` file.

The component starts by importing `React`, `useState`, and `useEffect` from the React library and `updateSpeaker` from the `SpeakersAPI.js` module. When the form is submitted, the `handleSubmit` function is called; it calls the `updateSpeaker` function from the `SpeakersAPI.js` file and passes in `speakerId` and an object containing the updated speaker's information. If the request is successful, it sets the success state to true and if there is an error, it sets the error state to `error.message`.

Now, you need to update the `SpeakersAPI.js` file in `src/services/SpeakersAPI.js` to add the `updateSpeaker` API call function:

```
// API function to update a speaker
export const updateSpeaker = (speakerId, speakerData) => {
    const url = `${API_URL}/speakers/${speakerId}`;
    return axios
        .put(url, speakerData, { headers: addHeaders() })
        .then((response) => response.data)
        .catch(handleErrors);
};
```

The preceding code defines an `updateSpeaker` API function used for updating a speaker's information on the backend:

- The function takes two parameters: `speakerId` (representing the ID of the speaker to be updated) and `speakerData` (an object containing the updated speaker information).

- It constructs the URL for the API endpoint by appending `speakerId` to the base URL.

- The function uses the Axios library to send a PUT request to the constructed URL, passing `speakerData` as the request payload and including the appropriate headers using the `addHeaders` function.

- If the request is successful, it returns the response data. If an error occurs during the request, it catches the error and calls the `handleErrors` function to handle and propagate the error.

Next, you will learn how to delete speaker data from the database.

Deleting data from a database – the React–Flask approach

Deleting data from a database involves removing one or more records or rows from a table. In this section, you are going to learn how to handle delete requests in a React–Flask web application.

Handling delete requests in Flask

Let's create the endpoint to handle the logic for deleting speaker data from the database:

```
@app.route('/api/v1/speakers/<int:speaker_id>',
    methods=['DELETE'])
def delete_speaker(speaker_id):
    speaker = Speaker.query.get_or_404(speaker_id)
    if not current_user.has_permission("delete_speaker"):
        abort(http.Forbidden("You do not have permission to
            delete this speaker"))
    events =
        Event.query.filter_by(speaker_id=speaker_id).all()
    if events:
        abort(http.Conflict("This speaker has associated
            events, please delete them first"))
    try:
        if speaker.speaker_avatar:
            speaker_avatar.delete(speaker.speaker_avatar)
        with db.session.begin():
            db.session.delete(speaker)
    except Exception:
        abort(http.InternalServerError("Error while
            deleting speaker"))
    return jsonify({"message": "Speaker deleted
        successfully"}), http.OK
```

The preceding code defines an API route for deleting a speaker, performs the necessary checks, deletes the speaker from the database, handles errors, and returns an appropriate response.

Next, we will explore the React component for handling delete requests from the frontend.

Handling delete requests in React

When building a React application, you can handle a delete request to remove a speaker resource by creating a component that interacts with the backend API. This component will send the delete request to the appropriate endpoint, handle any potential errors, and update the component's state accordingly to reflect the deletion of the speaker resource.

Let's start by creating a DeleteSpeaker component. In frontend/src/pages/Admin/Speakers/DeleteSpeaker.js, add the following code:

```
import React, { useState, useEffect } from "react";
import { useParams, useNavigate } from "react-router-dom";
```

```
import { deleteSpeaker } from "./api/SpeakersAPI";

const DeleteSpeaker = () => {
    const { speakerId } = useParams();
    const navigate = useNavigate();
    const [error, setError] = useState("");
    const [isLoading, setIsLoading] = useState(false);

    const handleDelete = async () => {
        try {
            setIsLoading(true);
            await deleteSpeaker(speakerId);
            setIsLoading(false);
            navigate("/speakers"); // Redirect to speakers
                                   list after successful
                                   deletion
        } catch (err) {
            setIsLoading(false);
            setError("Failed to delete speaker.");
        }
    };

    useEffect(() => {
        return () => {
            // Clear error message on component unmount
            setError("");
        };
    }, []);

    return (
        <div>
            {error && <p className="error">{error}</p>}
            <p>Are you sure you want to delete this
                speaker?</p>
            <button onClick={handleDelete}
                disabled={isLoading}>
                {isLoading ? "Deleting..." : "Delete
                Speaker"}
            </button>
        </div>
    );
};

export default DeleteSpeaker;
```

The preceding code defines a component that allows users to delete a speaker by id. The component starts by importing the useParams and useNavigate hooks from react-router-dom to extract the speakerId value from the URL. It also imports the deleteSpeaker function from src/services/SpeakersAPI.js to handle the deletion of the speaker with an API call on the backend. The component then uses the useState hook to initialize two state variables: error and success.

The component has a single button that, when clicked, triggers the handleDelete function. This function prevents the default form submission behavior and then calls the deleteSpeaker function passing in speakerId as an argument. If the deletion is successful, it sets the success state to true; otherwise, it sets the error state to the error message returned from the API. The component then renders a message to indicate whether the deletion was successful or there was an error.

Now, you need to update the SpeakersAPI.js file in src/api/SpeakersAPI.js to add the deleteSpeaker API call function:

```
// API function to delete a speaker
export const deleteSpeaker = async (speakerId) => {
    const url = `/api/v1/speakers/${speakerId}`;
    try {
        const speakerResponse = await axios.get(url);
        const speaker = speakerResponse.data;
        if (!speaker) {
            throw new Error("Speaker not found");
        }
        const eventsResponse = await
           axios.get(`/api/v1/events?speakerId=${speakerId}`
           );
        const events = eventsResponse.data;
        if (events.length > 0) {
          throw new Error("This speaker has associated
             events, please delete them first");
        }
        await axios.delete(url);
        return speaker;
    } catch (err) {
        if (err.response) {
            const { status, data } = err.response;
            throw new Error(`${status}: ${data.error}`);
        } else if (err.request) {
            throw new Error('Error: No response received
                from server');
        } else {
            throw new Error(err.message);
```

```
        }
    }
};
```

The preceding code defines a `deleteSpeaker` function that takes in a `speakerId` as its parameter. The function uses the Axios library to make HTTP requests to the server. The function starts by trying to get the speaker details from the server by making a `GET` request to the `/api/v1/speakers/{speakerId}` endpoint.

It then checks whether the speaker exists. If the speaker doesn't exist, the function throws an error with the message **Speaker not found**. If the speaker exists, the function makes a `GET` request to the `/api/v1/events?speakerId=${speakerId}` endpoint to get a list of events associated with the speaker. It then checks whether the length of events is greater than 0. If so, it throws an error with the message, **This speaker has associated events, please delete them first**.

Finally, the function makes a `DELETE` request to the `/api/v1/speakers/{speakerId}` endpoint to delete the speaker. If there's an error during the process, the function checks the error and throws an appropriate error message. The function then exports the `deleteSpeaker` function so it can be imported and used in other parts of the application.

Next, we will discuss how pagination can be handled in the React–Flask application.

Managing pagination in a React–Flask application

When working with a large dataset, it's important to implement pagination to make the large dataset more manageable for the user. Pagination is a technique used to divide a large set of data into smaller, more manageable chunks called **pages**. Each page contains a subset of the total data, allowing users to navigate through the data in a controlled manner.

Pagination provides a way to present large datasets efficiently, improves performance, and enhances the user experience by making data more accessible. In this section, you will learn how to implement pagination in a React–Flask web application. To implement pagination, you will need to make some changes to the backend server to handle pagination requests.

You can use the Flask-SQLAlchemy library to handle pagination on the backend. On the Flask backend, you can implement pagination for the `speaker` model using the Flask-SQLAlchemy library's pagination feature. Let's delve into how you can implement pagination for the `Speaker` model.

Update `get_speakers()` with the following code in the `app.py` file:

```python
from flask_sqlalchemy import Pagination
@app.route('/api/v1/speakers', methods=['GET'])
def get_speakers():
    page = request.args.get('page', 1, type=int)
    per_page = request.args.get('per_page', 10, type=int)
```

```
speakers = Speaker.query.paginate(page, per_page,
    False)
if not speakers.items:
    return jsonify({"error": "No speakers found"}), 404
return jsonify({
    'speakers': [speaker.serialize() for speaker in
        speakers.items],
    'total_pages': speakers.pages,
    'total_items': speakers.total
}), 200
```

In the preceding code, we are using the `paginate()` method from Flask-SQLAlchemy to add the pagination functionality to the speakers' collection. The `page` and `per_page` arguments are passed in as query parameters in the `GET` request. The default value for `page` is 1 and `per_page` is 10.

For the React frontend, you can use the `useState` and `useEffect` hooks to handle pagination in a functional component.

Let's modify the `ViewSpeakers` components and add pagination functionality to the component:

```
import React, { useState, useEffect } from 'react';
import { getSpeakers } from
    '../../../services/SpeakersAPI';

const ViewSpeakers = () => {
    const [speakers, setSpeakers] = useState([]);
    const [currentPage, setCurrentPage] = useState(1);
    const [speakersPerPage] = useState(10);
    const [isLoading, setIsLoading] = useState(false);
    const [error, setError] = useState(null);

    useEffect(() => {
        fetchSpeakers();
    }, []);

};


export default ViewSpeakers;
```

The preceding code defines a component that displays a list of speakers using pagination. The component makes use of React hooks to manage its state. The `speakers` state variable is used to store the list of speakers, and the `page` and `perPage` state variables are used to store the current page number and the number of items to be displayed per page, respectively.

The useEffect hook is used to fetch the speakers from the server when the component is mounted and whenever the page or perPage state variables change. The fetchSpeakers function uses the Axios library to make a GET request to the '/api/v1/speakers?page=${page}&per_page=${perPage}' endpoint, passing in the current page number and the number of items per page as query parameters.

The response data is then stored in the speakers state variable. The ViewSpeakers component then maps through the speakers' array and displays the name and email of each speaker. The component also includes two buttons, one for navigating to the previous page and one for navigating to the next page.

The onClick handlers of these buttons update the page state variable accordingly, and the **Previous** button is also disabled if the current page is 1 to prevent the user from navigating to a non-existent previous page.

Summary

In this chapter, we discussed in detail how you can fetch and display data in a React–Flask web application. We examined one of the ways fetching and displaying data is handled. You were able to work from the backend in defining the Speaker model class and implement various endpoints to handle data fetching from the database, and adding, updating, and deleting data on it.

We used the Axios library to send a request to the Flask backend, which then retrieved the data from a database and returned it to the frontend in a response. The React frontend then processed the response and displayed the data to the end user. Lastly, we implemented pagination as a way to present large datasets efficiently and to improve the performance of React–Flack web application projects.

Next, we are going to discuss authentication and authorization in a React–Flask application and examine the best practices to ensure that your application is secure and ready for production.

12

Authentication and Authorization

In building full-stack web applications, you will more than often want to implement a system that allows users to trust you with their sensitive information. As a full-stack web developer, it is crucial to understand how to implement robust authentication and authorization mechanisms. You need to know how to protect the security of user data and the integrity of your application. Imagine you are building an e-commerce website that allows users to make online purchases.

If you do not properly authenticate and authorize users, it would be possible for someone to gain unauthorized access to the website and place orders using someone else's personal information. This could result in financial loss for the legitimate user as well as damage to the reputation of an online business or that of your clients.

Furthermore, if you fail to properly authenticate and authorize users, it could also open your web application up to attacks such as SQL injection, where an attacker can gain access to sensitive information stored in your database. This could lead to the loss of customer data and legal repercussions may be staring you in the face.

In this chapter, we will delve into the world of web security and explore the best practices and techniques for securing Flask web applications. As the famous computer scientist Bruce Schneier once said, *Security is a process, not a product* (`https://www.schneier.com/essays/archives/2000/04/the_process_of_secur.html`). This chapter will equip you with the knowledge and skills needed to understand the importance of information security and how to implement it in a Flask application.

From understanding the fundamentals of authentication and authorization to managing user sessions and creating accounts with secure passwords, this chapter will cover the crucial elements of web application security. We will examine the process of securing your Flask application and show you how to implement these concepts in practice.

In this chapter, you will learn about the following topics:

- Understanding the fundamentals of information security
- Defining authentication and the authentication role in web applications
- Implementing password security and hashing passwords
- Understanding access and authorization in web application development
- Adding authentication to your Flask application
- Identifying system users and managing their information
- Session management
- Creating a password-protected dashboard
- Implementing flash messages in Flask

Technical requirements

The complete code for this chapter is available on GitHub at: `https://github.com/PacktPublishing/Full-Stack-Flask-and-React/tree/main/Chapter12`.

Understanding the fundamentals of information security

Information security is a critical aspect of web application development. In today's digital age, personal and sensitive information is often stored and transmitted through web applications, making them vulnerable to various types of security threats. These threats can range from simple attacks such as **SQL injection** and **cross-site scripting** (**XSS**) to more complex attacks such as **man-in-the-middle** (**MITM**) and **distributed denial of service** (**DDoS**).

Let's delve deeper into some of the various types of threats that can compromise your web application security:

- **SQL injection**: This is a type of cyberattack in which an attacker injects malicious SQL code into an application's input fields to trick the application into executing unintended database actions. This can lead to unauthorized access, data manipulation, or even data leakage.

  For instance, consider a web application login form where a user enters their `username` and `password` details. If the application is vulnerable to SQL injection, an attacker could input something like `' OR '1'='1` in the password field.

  The SQL query may then become `SELECT * FROM users WHERE username = 'username' AND password = '' OR '1'='1';`, which would potentially allow the attacker to log in without a valid password.

- **Cross-site scripting (XSS):** This is a vulnerability that allows attackers to inject malicious scripts into web pages viewed by other site users. These scripts can steal user information, manipulate page content, or perform other malicious actions.

  For instance, imagine your Flask application displays user comments on a web page without proper sanitization:

  ```
  comment = request.form['comment']
  html = f"<p>{comment}</p>"
  ```

  If an attacker submits a comment such as `<script>malicious_scripts()</script>`, other users viewing the comment section might execute the script unintentionally.

- **Cross-site request forgery (CSRF):** This is an attack where an attacker tricks a user into unknowingly making a request to a web application on which the user is authenticated. This can lead to unauthorized actions being taken on behalf of the user without their consent.

  CSRF attacks exploit the trust that a website has in a user's browser. For instance, an unsuspecting user logs into an online banking website and gets a session cookie. The attacker creates a malicious web page that contains a hidden form that submits a request to transfer money from the user's account to the attacker's account.

  The user visits the attacker's web page, and the hidden form is submitted using the user's session cookie, resulting in an unauthorized transfer. This type of attack exploits the trust that a website has in the user's browser to perform unauthorized actions.

- **Distributed Denial of Service (DDoS) attacks:** This type of attack involves overwhelming a target server, service, or network with a flood of traffic from multiple sources, rendering it inaccessible to legitimate users. For instance, an attacker might use a botnet (a network of compromised computers) to send a massive amount of traffic to a web application. This can cause a web application to become slow or entirely unavailable to users.

However, there are ways you can mitigate against these malicious threats, which are capable of undermining your web application. Now, we will highlight some of the best practices for securing your web application.

- **Input validation:** You need to ensure that all input data is properly sanitized and validated to prevent SQL injection and XSS attacks.

- **Use prepared statements:** Prepared statements allow you to create a database-independent API for executing SQL statements. Prepared statements can only be executed, not constructed, which makes it much harder for an attacker to inject malicious code into the database. ORM libraries, such as SQLAlchemy in Flask, handle the construction of SQL queries for you and provide a safe and efficient way to interact with databases.

- **Password storage:** Store passwords securely using a strong hashing algorithm and a unique salt for each user.

- **Use of HTTPS**: Use HTTPS to encrypt all communication between the client and the server to prevent eavesdropping and MITM attacks.

- **Session management**: Properly manage sessions to prevent session hijacking and fix session fixation vulnerabilities in your web application.

- **Access control**: Use role-based access control to restrict access to sensitive resources and functionality.

- **Logging and monitoring**: You need to consistently keep detailed logs of all application activity and monitor them for suspicious activity.

- **Using up-to-date software**: You need to regularly update the framework, libraries, and all dependencies that your web application is using to ensure that known vulnerabilities are patched.

- **Using security headers**: Use security headers such as `X-XSS-Protection`, `X-Frame-Options`, and `Content-Security-Policy` to prevent certain types of attacks.

- **Regularly testing for vulnerabilities**: Regularly conduct penetration testing and vulnerability scanning to identify and fix any security issues.

In the remaining parts of this chapter, we will discuss and implement authentication with authorization in a Flask web application to help you keep your application and its users' data secure.

Next, we will discuss authentication and the authentication role in web applications. This will improve your understanding of how to verify users' identities and the various types of authentications.

Defining authentication and the authentication role in web application

Authentication is the process of verifying the identity of a user and ensuring that only authorized users have access to the application's resources and functionality. Authentication is an important aspect of any web application, including those built with Flask.

This is typically done by prompting the user to provide a set of credentials, such as a username and password, that the web application can use to confirm the user's identity. The purpose of authentication in web application development is to ensure that only authorized users can access sensitive information and perform certain actions within a web application.

In web development, we have several types of authentication methods that can be used in any web application project. These are some of the most commonly used methods:

- **Password-based authentication**: This is the most common form of authentication we encounter in everyday use and involves the user entering a username/email and password to gain access to the web application. This method is simple and easy to implement but comes with its weakness. Password-based authentication is vulnerable to attacks such as brute-force and dictionary attacks.

- **Multi-factor authentication (MFA):** This method adds an additional layer of security by requiring the user to provide multiple forms of identification. For instance, a user may be required to enter a password and also provide a one-time code that's been sent to their phone or email. MFA is more secure than password-based authentication but can negatively impact the user experience.

- **Token-based authentication:** This method involves the user being issued a token that they must present to the web application to gain access. Tokens can be in the form of a JWT or OAuth token and are often stored in a browser's cookies or local storage. Tokens can easily be revoked, making it easier to maintain security.

- **Biometric authentication:** This method involves the use of biological characteristics such as fingerprints, facial recognition, or voice recognition to verify a user's identity. Biometric authentication is considered to be more secure than other methods but can be more expensive to implement.

When you are deciding which authentication method to use, it's crucial to consider the level of security required for the web application and the user experience. Each of these authentication methods has its pros and cons. It is essential to choose the right method for your application.

For instance, if you are building a web application that requires a high level of security, you may want to consider using MFA or biometric authentication. And of course, biometric authentication is rarely used in public or general-purpose web applications. If you are building a simple web application that does not require a high level of security, password-based authentication may be safe and sufficient.

Next, we will discuss the concept of implementing password security and hashing passwords in securing a Flask web application.

Implementing password security and hashing passwords

In any web application that requires access, passwords are often the first line of defense against unauthorized access. As a developer, you will want to ensure that passwords are securely managed when building Flask applications. A critical component of password management in web applications is to never store passwords in plaintext.

Instead, passwords should be hashed, which is a one-way encryption process that produces a fixed-length output that cannot be reversed. When a user enters their password, it is hashed and compared with the stored hash. If the two hashes match, the password is correct. Hashing passwords can help protect against attacks such as brute-force and dictionary attacks.

Brute-force attacks involve trying every possible combination of characters to find a match, while dictionary attacks involve trying a pre-computed list of words. Hashing passwords makes it computationally infeasible for an attacker to reverse the hash and discover the original password.

In Flask, you can use a library such as `Flask-Bcrypt` to handle password hashing. `Flask-Bcrypt` is a Flask extension that provides `bcrypt` password hashing for Flask. `Flask-Bcrypt` provides a simple interface for hashing and checking passwords. You can also use `Flask-Bcrypt` to generate random salts for use in password hashing.

Let's quickly run through an example of password hashing using `Flask-Bcrypt`:

```
from flask import Flask, render_template, request
from flask_bcrypt import Bcrypt
app = Flask(__name__)
bcrypt = Bcrypt()
@app.route("/", methods=["GET", "POST"])
def index():
    if request.method == "POST":
        password = request.form.get("password")
        password_hash =
            bcrypt.generate_password_hash(password)
                .decode('utf-8')
        return render_template("index.html",
            password_hash=password_hash)
    else:
        return render_template("index.html")
@app.route("/login", methods=["POST"])
def login():
    password = request.form.get("password")
    password_hash = request.form.get("password_hash")
//Check GitHub for the complete code

if __name__ == "__main__":
    app.run(debug=True)
```

The preceding code uses the `Flask Bcrypt` library to hash and check a password. It imports the `Bcrypt` class and the `check_password_hash` function, creating an instance of `Bcrypt` with the Flask application. When the form is submitted, the password is hashed using the `flask_bcrypt` extension, and the hashed password is displayed back to the user on the same page. The `render_template` function is used to render the HTML templates, and the `Bcrypt` extension is used for secure password hashing.

Next, we will discuss access and authorization in web application development.

Understanding access and authorization in web application development

Access and authorization in web application development is the process of controlling who has access to specific resources and actions within a web application. As a developer, you will want to design and ensure that users can only perform actions they are authorized to perform and access resources they are authorized to access in a web application.

As discussed earlier, authentication is the process of verifying the identity of a user. Authorization is the process of determining what a user is allowed to do within a web application. When you combine these two mechanisms, you have a system that ensures that only authorized users can access sensitive information and perform certain actions within a web application.

Several different types of access control methods can be used in web application development. We will discuss some of them and make specific reference to how Flask can handle access and authorization:

- **Login and password systems**: Login and password systems can restrict access to certain resources and actions based on the user's authentication status. For instance, only logged-in users may be allowed to access certain pages, while anonymous users may be redirected to the login page. In a Flask application, a login and password system can be implemented using extensions such as `Flask-Login` and `Flask-Security`.

- **Open Authorization (OAuth)**: OAuth is a standard for authorization that allows users to grant third-party applications access to their resources without them having to share their credentials, such as their username and password. Rather, the user grants permission to the third-party application. This third-party application can then access the user's resources on the user's behalf.

   An example of an OAuth implementation is social media logins when a user logs in with a Google or Facebook account. When a user clicks the **Log in with Google** or **Log in with Facebook** button, the user is redirected to the relevant provider's website to grant permission. Once the user grants permission, the provider sends an authorization code back to the application, which can then exchange the code for an access token.

   This access token can then be used to make API calls and access the user's resources. In a Flask application, OAuth can be implemented using extensions such as `Flask-OAuthlib`. This extension provides support for `OAuth 1.0a` and `OAuth 2.0`. `Flask-OAuthlib` makes it easy for developers to implement OAuth in their Flask applications.

- **JSON Web Token (JWT)**: JWT is a widely used standard for authentication and authorization in web applications. JWT provides a secure way to transmit information between parties as a JSON object that is signed by the server and can be verified by the client. In a JWT-based authentication system, the server generates a JWT when the user logs in and sends it to the client.

The client then sends the JWT back to the server with each subsequent request to authenticate the user. The server verifies the signature of the JWT. In a Flask application, JWT-based authentication can be implemented using extensions such as `Flask-JWT` and `Flask-JWT-Extended`.

These extensions provide features such as token generation, verification, and expiration, as well as the ability to restrict access to certain resources and actions based on the claims contained in the JWT to ensure that it was generated by a trusted source and has not been tampered with.

- **Role-based access control** (**RBAC**): This is a method of regulating access to resources or actions based on the roles that a user holds. In an RBAC system, users are assigned roles, and each role is associated with a set of permissions that determine what actions the user is allowed to perform.

  For instance, in a web application, an administrator may have permission to create and edit user accounts, while a regular user may only have permission to view their own account information. In a Flask application, RBAC can be implemented using extensions such as `Flask-RBAC`.

  The `Flask-RBAC` extension provides features such as role management, permission management, and the ability to restrict access to certain resources and actions based on the user's role.

- **Policy-based access control** (**PBAC**): This is a method of regulating access to resources or actions based on predefined policies. In the PBAC system, policies are defined to specify the conditions under which access is granted or denied. For instance, in a web application, a policy may be defined to grant access to a resource only if the user is authenticated and their account is active.

  The policy may also specify that access should be denied if the user has reached a certain number of failed login attempts. In a Flask application, PBAC can be implemented using extensions such as `Flask-Policies`. `Flask-Policies` provides features such as policy management, enforcement, and the ability to restrict access to certain resources and actions based on the conditions specified in the policies.

By using these libraries, you can easily handle user roles and permissions and restrict access to certain views and routes based on the user's role. Next, we will take a look at how to implement authentication in a Flask web application.

Adding authentication to your Flask application

JWT is a popular method for authentication in modern web applications. A JWT is a JSON object that is digitally signed and can be used to authenticate users by transmitting claims between parties, such as an authorization server and a resource server. In a Flask web application, you can use the `PyJWT` library to encode and decode JWTs for authentication.

When a user logs into a Flask application, the backend verifies the user's credentials, such as their email and password, and if they are valid, a JWT is generated and sent back to the client. The client stores the JWT in the browser's local storage or as a cookie. For subsequent requests to protected routes and resources, the client sends the JWT in the request header.

The backend decodes the JWT to verify the user's identity, grants or denies access to the requested resources, and generates a new JWT for subsequent requests. JWT for authentication allows stateless authentication. This means that the authentication information is stored in the JWT, which can be passed around between different servers, instead of on the server's memory. This makes it easier to scale the application and reduces the risk of data loss or corruption.

JWT authentication also enhances security by using digital signatures to prevent data tampering. The signature is generated using a secret key that's shared between the server and the client. The signature ensures that the data in the JWT has not been altered in transit. JWT authentication is a secure and efficient method for authenticating users in a Flask application.

By implementing JWT authentication in a Flask application, developers can simplify the process of authenticating users and reduce the risk of security breaches. Let's examine the backend and frontend implementation of JWT.

Flask backend

The following code defines two Flask endpoints – /api/v1/login and /api/v1/dashboard:

```
@app.route('/api/v1/login', methods=['POST'])
def login():
    email = request.json.get('email', None)
    password = request.json.get('password', None)
    if email is None or password is None:
        return jsonify({'message': 'Missing email or
            password'}), 400
    user = User.query.filter_by(email=email).first()
    if user is None or not bcrypt.check_password_hash
        (user.password, password):
        return jsonify({'message': 'Invalid email or
            password'}), 401

    access_token = create_access_token(identity=user.id)
    return jsonify({'access_token': access_token}), 200
@app.route('/api/v1/dashboard', methods=['GET'])
@jwt_required
def dashboard():
    current_user = get_jwt_identity()
    user = User.query.filter_by(id=current_user).first()
    return jsonify({'email': user.email}), 200
```

The /api/v1/login endpoint is for handling user login requests. It takes in a JSON request with two properties: email and password. If either of these properties is missing, the function returns a JSON response with a message indicating Missing email or password and a status code of 400 (Bad Request).

Next, the function queries the database for a user with the given email. If no such user exists, or if the password provided does not match the hashed password stored in the database, the function returns a JSON response with a message indicating Invalid email or password and a status code of 401 (Unauthorized).

Otherwise, the function generates a JWT using the create_access_token function and returns it as a JSON response, along with a status code of 200 (OK). The JWT can be used to authenticate the user in subsequent requests to the backend. The /api/v1/dashboard endpoint is a protected endpoint that can only be accessed by users who have a valid JWT.

The jwt_required decorator is used to enforce this restriction. When this endpoint is accessed, the JWT is used to extract the user's identity, which is then used to retrieve the user's email from the database. This email is then returned as a JSON response, along with a status code of 200 (OK).

React frontend

The following code shows a login form and a dashboard. The LoginForm component has three states – email, password, and accessToken. When the form is submitted, it makes a POST request to the /api/v1/login endpoint with the email and password data, and the response of the request is stored in the accessToken state:

```
import React, { useState } from 'react';
import axios from 'axios';
const LoginForm = () => {
  const [email, setEmail] = useState('');
  const [password, setPassword] = useState('');
  const [accessToken, setAccessToken] = useState('');
  const handleSubmit = async (event) => {
    event.preventDefault();
    try {
      const res = await axios.post('/api/v1/login', {
        email, password });
      setAccessToken(res.data.access_token);
    } catch (err) {
      console.error(err);
    }
  };
```

```
    return (
      <>
        {accessToken ? (
          <Dashboard accessToken={accessToken} />
        ) : (
          <form onSubmit={handleSubmit}>
            ....
            />
            <button type="submit">Login</button>
          </form>
        )}
      </>
    );
  };

};
  export default LoginForm;
```

The Dashboard component takes an accessToken prop and has one state, email. It makes a GET request to the /api/v1/dashboard endpoint with an authorization header set to accessToken, and the response is stored in the email state. The component displays a message stating "Welcome to dashboard, [email]!".

The LoginForm component returns either the Dashboard component if accessToken is truthy, or the login form if accessToken is falsy.

Next, we will discuss how to identify web application users and manage their information.

Identifying system users and managing their information

In most web applications, users are identified by a unique identifier such as a username or email address. Typically, in a Flask application, you can use a database to store user information, such as usernames, email addresses, and hashed passwords.

When a user attempts to log in, the entered credentials (username and password) are compared to the information stored in the database. If the entered credentials match, the user is authenticated, and a session is created for that user. In Flask, you can use the built-in session object to store and retrieve user information.

By using sessions, you can easily identify users in a Flask web application and retrieve information about them. However, it's important to note that sessions are vulnerable to session hijacking attacks. So, it's essential to use secure session management techniques such as regenerating session IDs after login and using secure cookies.

Let's examine an implementation example:

```python
from flask import Flask, request, redirect, session, jsonify
app = Flask(__name__)
app.secret_key = 'secret_key'
@app.route('/login', methods=['POST'])
def login():
    data = request.get_json()
    email = data.get('email')
    password = data.get('password')
    session['email'] = email
    return jsonify({'message': 'Login successful'}), 201

@app.route('/dashboard', methods=['GET'])
def dashboard():
    email = session.get('email')
    user = User.query.filter_by(email=email).first()
    return jsonify({'email': email, 'user':
        user.to_dict()}), 200
```

In the preceding code, the first line imports the required modules from the Flask library. The next line creates an instance of the `Flask` class and assigns it to the `app` variable. The `app.secret_key` attribute is set to `'secret_key'`, which is used to securely sign the session cookie.

The login function is defined as a POST endpoint at the `api/v1/login` route. This endpoint uses the `request.get_json()` method to get the JSON data from the request body and extract the values for `email` and `password`. `email` is then stored in the session using `session['email'] = email`. The function returns a JSON response with a message of `"Login successful"` and a `201` status code, indicating the successful creation of a resource.

Then, the dashboard function is defined as a GET endpoint at the `api/v1/dashboard` route. It retrieves `email` from the session using `session.get('email')`. The function then queries the database for a user with the specified email using `User.query.filter_by(email=email).first()`. The `email` and user data (converted into a dictionary using `to_dict()`) are returned in a JSON response with a 200 status code, indicating the successful retrieval of a resource.

You can also identify users in a Flask application with a token-based authentication method. In this method, a token is issued to the user when they log in, and the token is stored in the user's browser as a cookie or placed in local storage. This token is then sent with each subsequent request made by the user, and the server uses this token to identify the user. JWT is a commonly used token format, and libraries such as `Flask-JWT` and `Flask-JWT-Extended` make it easy to implement JWT-based authentication in Flask.

Next, we will delve deeper into tracking a user's session in a web application.

Session management

Session management is a critical aspect of web development as it enables a web application to identify and track a user's actions over a certain period. In Flask web applications, session management is typically implemented on the server side using Flask's in-built session object or a Flask extension such as `Flask-Session`; on the frontend React side, you can use React's `localStorage` or `sessionStorage`.

Flask thrives on its simplicity as a framework of choice for Python that makes it easy to build small to enterprise-sized web applications. Flask can manage user sessions using the built-in session object and some of the available Flask extensions contributed by the community members.

A session object is a dictionary-like object that is stored on the server and can be accessed by the client via a secure session cookie. To use a session object, a *secret key* must be set in the Flask application. This secret key is used to encrypt and sign the session data, which is stored in a secure cookie on the client's browser. When a user visits a protected resource, the server verifies the session cookie and grants access if the cookie is valid.

Let's implement session management in a Flask backend and React frontend. We will create a counter endpoint that keeps track of the number of times a user visited a dashboard page.

Flask backend

We will use `Flask-Session` to store session data and securely manage sessions. To use `Flask-Session`, you need to install it first. You can do this by running the `pip install flask-session` command in the Terminal.

Once you've installed `Flask-Session`, you need to add the following code to your Flask application:

```
from flask import Flask, session
from flask_session import Session
app = Flask(__name__)
app.config["SESSION_TYPE"] = "filesystem"
Session(app)
@app.route("/api/v1/couters")
def visit_couter():
    session["counter"] = session.get("counter", 0) + 1
    return "Hey , you have visited this page:
        {}".format(session["counter"])
```

The preceding code shows a simple implementation of session management in a Flask backend:

1. The first line imports the Flask module, while the second line imports the `Flask-Session` extension.

2. The next few lines create a Flask application object and configure the session type to be stored on the filesystem.

3. The `Session` object is then initialized with the Flask application object as its argument.

4. The `@app.route` decorator creates a route – in this case, `/api/v1/counters` – for the `visit_counter` function.

5. The `visit_counter` function retrieves the current value of the `counter` key in the session or sets it to `0` if it doesn't exist, and then increments the value by `1`. The updated value is then returned to the user in the response.

Let's explore the React frontend part of this implementation.

React Frontend

You can use the Axios library to send HTTP requests to the Flask server. If not installed yet, you can install Axios with the `npm install axios` command.

Once you've installed Axios, you can use it to send an HTTP request to the Flask server to set or get the session data:

```
import React, { useState } from "react";
import axios from "axios";
function VisitCouter() {
    const [counter, setCounter] = useState(0);
    const getCounter = async () => {
        const response = await axios.get(
            "http://localhost:5000/api/v1/counters");
        setCounter(response.data.counter);
        };
        return (
          <div>
            <h1>You have visited this page: {counter}
              times!</h1>
            <button onClick={getCounter}>Get Counter
              </button>
          </div>
        );
}
export default VisitCounter;
```

The preceding code demonstrates the frontend implementation of a React frontend that retrieves the visit counter from a Flask backend:

1. The first line imports the required libraries – that is, `React` and `axios`.

2. The next section declares the `VisitCounter` function component, which returns a view for the user.

3. Within the component, the state variable counter is initialized using the `useState` hook.

4. The `getCounter` function uses the `axios` library to make a GET request to the `/api/v1/counters` endpoint on the Flask backend. The response from the backend, which contains the updated counter value, is then used to update the counter state variable.

5. The component returns a div that displays the value of the counter and a button that, when clicked, triggers the `getCounter` function to retrieve the updated counter value from the backend.

Next, we will discuss how to create a password-protected dashboard in a Flask-React web application.

Creating a password-protected dashboard

Protecting pages in a web application is essential for maintaining security and privacy. By extension, this can help prevent unauthorized access to sensitive information. In this section, you will be implementing a protected dashboard page in a Flask-React web application.

A dashboard is a user-friendly interface that provides an overview of data and information. The data that's displayed on a dashboard can come from a variety of sources, such as databases, spreadsheets, and APIs.

Flask backend

The following code demonstrates an implementation that allows an admin user to log in and see a protected dashboard page. We will implement minimalist login and logout endpoints that define login and logout functionality and protect the `dashboard` endpoint. The application uses the `Flask-Session` library to store session data in the filesystem:

```
from flask import Flask, request, jsonify, session
from flask_session import Session
app = Flask(__name__)
app.config["SESSION_TYPE"] = "filesystem"
Session(app)
@app.route("/api/v1/login", methods=["POST"])
def login():
    username = request.json.get("username")
    password = request.json.get("password")
    if username == "admin" and password == "secret":
        session["logged_in"] = True
        return jsonify({"message": "Login successful"})
    else:
        return jsonify({"message": "Login failed"}), 401
@app.route("/api/v1/logout")
def logout():
    session.pop("logged_in", None)
```

```
        return jsonify({"message": "Logout successful"})
@app.route("/api/v1/dashboard")
def dashboard():
    if "logged_in" not in session:
        return jsonify({"message": "Unauthorized access"}),
            401
    else:
        return jsonify({"message": "Welcome to the
            dashboard"})
```

In the `login` endpoint, the application receives a POST request with the `username` and `password` parameters in the request body in JSON format. The code checks if the `username` and `password` parameters match the predefined values – that is, `admin` and `secret`. If the values match, the code sets the `logged_in` key in the session data to `True`, indicating that the user is logged in.

It returns a JSON response with a message stating `Login successful`. If the values don't match, the code returns a JSON response with a message stating `Login failed` and a `401` HTTP status code, indicating unauthorized access.

The `logout` endpoint removes the `logged_in` key from the session data, indicating that the user is logged out. It returns a JSON response with a message stating `Logout successful`.

The dashboard endpoint checks if the `logged_in` key exists in the session data. If it does not, the code returns a JSON response with a message stating `Unauthorized access` and a `401` HTTP status code. If the `logged_in` key exists, the code returns a JSON response with a message stating `"Welcome to the dashboard"`.

React frontend

The following code snippet is a React component that displays a dashboard for a user. It uses React hooks, specifically `useState` and `useEffect`, to manage its state and update the user interface:

```
import React, { useState, useEffect } from "react";
import axios from "axios";
function Dashboard() {
  const [isLoggedIn, setIsLoggedIn] = useState(false);
  const [message, setMessage] = useState("");
  const checkLogin = async () => {
    const response = await axios.get(
      "http://localhost:5000/api/v1/dashboard");
    if (response.status === 200) {
      setIsLoggedIn(true);
      setMessage(response.data.message);
    }
  };
```

```
  useEffect(() => {
  checkLogin();
  }, []);

  if (!isLoggedIn) {
    return <h1>Unauthorized access</h1>;
  }
  return <h1>{message}</h1>;
}
export default Dashboard;
```

When the component is rendered, it makes an HTTP GET request to `http://localhost:5000/api/v1/dashboard` using the `axios` library. This is done in the `checkLogin` function, which is called by the `useEffect` hook when the component is mounted.

If the response from the server is `200 OK`, this means that the user is authorized to access the dashboard. The component's state is updated to reflect this by setting `isLoggedIn` to `true` and `message` to the message returned from the server. If the response is not `200 OK`, this means the user is unauthorized and `isLoggedIn` remains `false`.

Finally, the component returns a message that tells the user whether they have access to the dashboard. If `isLoggedIn` is `false`, it returns `Unauthorized access`. If `isLoggedIn` is `true`, it returns the message from the server.

In this way, you can create a password-protected dashboard that is only accessible to authenticated users using React and Flask with added security for your application.

Next, you will learn how to implement flash messages in Flask and React web applications.

Implementing flash messages in Flask

Flash messages enhance the user experience in any web application, providing informative and timely feedback to users. Flash is used to display status or error messages on web pages after a redirect. For instance, after a successful form submission, a message can be stored in the flash to display a success message on the redirected page.

The flash message is stored in the user's session, which is a dictionary-like object that can store information between requests. With flash messages, you can pass information between requests securely and efficiently. This is useful for displaying messages that don't need to persist for a long time or that need to be shown only once, such as success or error messages. Since flash messages are stored in the user's session, they are only accessible by the server and are not sent to the client in plain text, making them secure.

Let's modify the login and logout endpoints to show flash messages.

Flask backend

The following code demonstrates the implementation of a flash message system with login and logout endpoints. The code starts by importing the necessary modules and creating a Flask application. The `app.secret_key = "secret_key"` line sets the secret key, which is used to encrypt the flash messages stored in the session:

```python
from flask import Flask, request, jsonify, session, flash
from flask_session import Session
app = Flask(__name__)
app.config["SESSION_TYPE"] = "filesystem"
app.secret_key = "secret_key"
Session(app)

@app.route("/api/v1/login", methods=["POST"])
def login():
    username = request.json.get("username")
    password = request.json.get("password")

    if username == "admin" and password == "secret":
        session["logged_in"] = True
        flash("Login successful")
        return jsonify({"message": "Login successful"})
    else:
        flash("Login failed")
        return jsonify({"message": "Login failed"}), 401

@app.route("/api/v1/logout")
def logout():
    session.pop("logged_in", None)
    flash("Logout successful")
    return jsonify({"message": "Logout successful"})
```

The `login` endpoint is defined by the `login` function, which is bound to the `/api/v1/login` URL. The function retrieves the `username` and `password` values from the JSON data in the request, and checks if they match the predefined values of `"admin"` and `"secret"`. If the values match, the user's session is marked as logged in by setting the `logged_in` key in the session, and a flash message is set to indicate that the login was successful.

The function then returns a JSON response, indicating the login was successful. If the values do not match, a flash message is set, indicating login failed, and a JSON response indicating the login failure is returned. The logout endpoint is defined by the `logout` function, which is bound to the `/api/v1/logout` URL.

The function removes the `logged_in` key from the session, indicating that the user is no longer logged in, and sets a flash message indicating that the logout was successful. A JSON response indicating the logout was successful is then returned.

React frontend

The following snippet demonstrates a React functional component that represents the dashboard of a web application handling flash messages from the backend. The `Dashboard` component makes use of `useState` and `useEffect` hooks:

```
import React, { useState, useEffect } from "react";
import axios from "axios";
function Dashboard() {
    const [isLoggedIn, setIsLoggedIn] = useState(false);
    const [message, setMessage] = useState("");
    const [flashMessage, setFlashMessage] = useState("");

    const checkLogin = async () => {
        const response = await axios.get(
            "http://localhost:5000/api/v1/dashboard");
        if (response.status === 200) {
            setIsLoggedIn(true);
            setMessage(response.data.message);
        }
    };
                    .....
        if (!isLoggedIn) {
            return (
                <div>
                    <h1>Unauthorized access</h1>
                    <h2>{flashMessage}</h2>
                    <button onClick={() =>
                        handleLogin("admin", "secret")}>
                        Login</button>
```

The `Dashboard` component keeps track of the following state variables:

- `isLoggedIn`: A Boolean value indicating if the user is logged in or not. It is initially set to `false`.

- `message`: A string value that represents a message that is displayed on the dashboard.

- `flashMessage`: A string value that represents a flash message that is displayed on the page.

The Dashboard component has three functions:

- checkLogin: An asynchronous function that makes a GET request to the /api/v1/ dashboard endpoint to check if the user is logged in or not. If the response status is 200, it updates the isLoggedIn state variables to true and messages a value of response. data.message.

- handleLogin: An asynchronous function that makes a POST request to the /api/v1/ login endpoint with the provided username and password values as the request body. If the response status is 200, it updates the isLoggedIn state variables to true and flashMessage to the value of response.data.message. If the response status is not 200, it updates flashMessage to the value of response.data.message.

- handleLogout: An asynchronous function that makes a GET request to the /api/v1/ logout endpoint. If the response status is 200, it updates the isLoggedIn state variables to false and flashMessage to the value of response.data.message.

The useEffect hook is used to call the checkLogin function when the component is mounted.

Finally, the component returns a UI, depending on the value of isLoggedIn:. If the user is not logged in, it displays a message saying "Unauthorized access" and a **Login** button. If the user is logged in, it displays a message value of "Login successful".

In this way, you can use flash messages to provide feedback to the user in a React application via the frontend, and then use the Flask backend to enhance the user's experience. Overall, flash messages make web applications more interactive and user-friendly.

Summary

This chapter has provided a comprehensive overview of the fundamentals of information security and how to secure a Flask web application using authentication and authorization. You learned about the best practices and were provided with use cases for implementing authentication and authorization in a Flask application. We also discussed different types of authentication methods and access control methods.

You explored how to manage user sessions and implement password-protected dashboards. Additionally, this chapter has shown you how to use flash messages to provide feedback to users of web applications. You are expected to have garnered a solid understanding of how to secure a Flask application and be able to implement authentication and authorization in your projects.

In the next chapter, we will discuss how to handle errors in Flask web applications with React handling the frontend part of it. We will delve into in-built Flask debugging capabilities and learn how to handle custom error messages in React-Flask applications.

13
Error Handling

Error handling is a critical component in the user experience of any web application. **Flask** provides several built-in tools and options for handling errors in a clean and efficient manner. The goal of error handling is to catch and respond to errors that may occur during the execution of your application such as runtime errors, exceptions, and invalid user inputs.

Flask provides a built-in debugger that can be used to catch and diagnose errors during development. So, why is the concept of error handling so important in any web application? An error-handling mechanism provides meaningful error messages to users when things go south when expected to go north, helping to maintain the overall quality of the user experience. Also, proactive error handling makes debugging easy.

If error-handling implementation is well thought out, then debugging issues and identifying the root causes of problems in the application becomes easier. As a developer, you would also want to increase the reliability of your application by anticipating and handling potential errors. This invariably makes your application more reliable and less likely to break under unexpected conditions.

In this chapter, we will explore the different strategies and techniques for handling errors in Flask web applications. You will understand and learn how to use the built-in **Flask debugger**, implement **error handlers**, and create custom **error pages** in order to provide meaningful feedback to the user.

In this chapter, you will learn about the following topics:

- Using the Flask debugger
- Creating error handlers
- Creating custom error pages
- Tracking events in your application
- Sending error emails to administrators

Technical requirements

The complete code for this chapter is available on GitHub at: `https://github.com/PacktPublishing/Full-Stack-Flask-and-React/tree/main/Chapter13`.

Using the Flask debugger

Flask as a lightweight Python web framework is widely used for building web applications. One of the out-of-the-box benefits of using Flask is its built-in debugger, which provides a powerful tool for identifying and fixing errors in your application.

When an error occurs in your Flask application, the debugger will automatically be activated. The debugger will provide detailed information about the error, including a stack trace, source code context, and any variables that were in scope at the time the error occurred. This information is golden for determining the root cause of the error and possible ideas for fixing it.

The Flask debugger also provides some interactive tools that can be used to inspect the state of your application and understand what is happening. For instance, you can evaluate expressions and examine the values of variables. You can also set breakpoints in your code, and step through your code line by line to see how it is executed.

Let's examine this code snippet for illustration:

```
import pdb
@app.route("/api/v1/debugging")
def debug():
    a = 10
    b = 20
    pdb.set_trace()
    c = a + b
    return f"The result is: {c}"
```

In this instance, you can set a breakpoint at the line before `c = a + b`, as done in the preceding code, and run the application. When the breakpoint is hit, you can enter the debugger and inspect the values of a, b, and c. You can also evaluate expressions and see their results. For instance, to evaluate the expression a + b, you can type a + b in the debugger's command prompt and hit *Enter*. The result, 30, will be displayed. You can also step through your code line by line by using the n command to go to the next line, and the c command to continue execution until the next breakpoint.

In this way, you can use the Flask debugger's interactive tools to understand what is happening in your application and debug it more effectively. This can be especially useful when dealing with large or complex code bases. The Flask debugger's interactive tools are useful when it is difficult to understand what is causing an error without additional tools and information.

Aside from interactive tools, Flask also provides a debug mode that can be enabled to provide more detailed error messages. When the debug mode is enabled, Flask will display detailed error pages with information about the error including a stack trace and the source code context. This information can be extremely helpful for debugging complex issues.

To enable the Flask debugger, simply set the `debug` configuration value to `True` in your Flask application. In this book project, we set this parameter in the `.env` file. You should only use this in development, as it can reveal sensitive information about your application to anyone who has access to it.

Additionally, Flask allows third-party extensions that can be used to enhance the debugging experience. For instance, `Flask-DebugToolbar` provides a toolbar that can be added to your application to display information about the current request and its context.

Flask's built-in debugger is a powerful tool that can help you quickly identify and fix errors in your application. Whether you are working on a small project or an enterprise-grade application, the debugger provides valuable information that can help you resolve issues and improve the reliability and performance of your application.

Next, we will discuss and implement error handlers in Flask web applications.

Creating error handlers

Flask also provides a mechanism for handling errors called error handlers. Error handlers are functions that are invoked when a specific error occurs in your application. These functions can be used to return custom error pages, log information about the error, or perform any other action that is appropriate for the error. To define an error handler in the Flask web application, you need to use the `errorhandler` decorator.

The decorator takes the error code as its argument, and the function that it decorates is the error handler that will be invoked when that error occurs. The error handler function takes an error object as its argument, which provides information about the error that occurred. This information can be used to provide a more detailed error response to the client or to log additional information about the error for debugging purposes.

In Flask backend and **React** frontend applications, error handling is a crucial step in ensuring a smooth user experience. As mentioned earlier, the goal of error handlers is to provide meaningful feedback to the user when something goes wrong, rather than simply returning a generic error message.

For instance, you can define error handlers for errors `400`, `404`, and `500`.

Flask backend

The following code shows error handlers that are created for the HTTP error codes 404 (not found), 400 (bad request), and 500 (internal server error):

```python
from flask import jsonify
@app.errorhandler(404)
def not_found(error):
    return jsonify({'error': 'Not found'}), 404


@app.errorhandler(400)
def bad_request(error):
    return jsonify({'error': 'Bad request'}), 400


@app.errorhandler(500)
def internal_server_error(error):
    return jsonify({'error': 'internal server error'}), 500
```

The not_found, bad_request, and internal_server_error functions return a JSON response containing an error message, along with the corresponding HTTP error codes.

React frontend

In the React frontend, you can handle these errors by making an HTTP request to the Flask backend and checking the response for errors. For example, you can use **Axios** in React:

```javascript
import React, { useState, useEffect } from 'react';
import axios from 'axios';
const Speakers = () => {
    const [error, setError] = useState(null);
    useEffect(() => {
        axios.get('/api/v1/speakers')
        .then(response => {
            // handle success
        })
        .catch(error => {
            switch (error.response.status) {
                case 404:
                    setError('Resource not found.');
                    break;
                case 400:
                    setError('Bad request');
                    break;
                case 500:
```

```
                    setError('An internal server error
                        occurred.');
                    break;
                default:
                    setError('An unexpected error
                        occurred.');
                    break;
            }
        });
    }, []);
    return (
        <div>
            {error ? <p>{error}</p> : <p>No error</p>}
        </div>
    );
};
export default Speakers;
```

The preceding error-handling code illustrates a React frontend communicating with a Flask backend API. The code imports `React`, `useState`, and `useEffect` hooks, as well as the `axios` library for making API requests. The code then defines a functional `Speakers` component that makes an API GET request to the `/api/v1/speakers` endpoint at the backend.

The `useEffect` hook is used to manage the API call, and the response is handled in a `.then()` block for success or a `.catch()` block for errors. In the `.catch()` block, the error response status is checked and a specific error message is set based on the status code. For instance, if the status code is `404`, `Resource not found` will be set as the error.

The error message is then displayed in the UI using conditional rendering, with the `No error` text being displayed if there is no error. The error message is stored in the state using the `useState` hook, with the initial value being `null`.

Next, we will discuss and implement custom error pages in Flask web applications.

Creating custom error pages

In addition to error handlers in Flask, you can also create custom error pages that provide a better user experience. When an error occurs in your application, the error handler can return a custom error page with information about the error, instructions for resolving the issue, or any other content that may be appropriate.

To create a custom error page in Flask, simply create an error handler as described in the preceding section and return a JSON response that contains the content for the error page.

For instance, let's take a look at the JSON response containing a custom error message in the following code:

```
@app.errorhandler(404)
def not_found(error):
    return jsonify({'error': 'Not found'}), 404
```

The preceding code returns a JSON response containing an error message, along with the corresponding HTTP error codes, when a 404 error occurs. Let's define the React frontend to handle the UI with an ErrorPage component:

```
import React from 'react';
const ErrorPage = ({ error }) => (
    <div>
        <h1>An error has occurred</h1>
        <p>{error}</p>
    </div>
);
export default ErrorPage;
```

The preceding code shows the ErrorPage component that takes an error prop and displays it in the error message. You can use this component in your application to display the custom error page whenever an error occurs.

You can simply add the ErrorPage component to the rest of the application. For instance, use the following code to add the ErrorPage component to the Speaker component:

```
import React, { useState, useEffect } from 'react';
import axios from 'axios';
import ErrorPage from './ErrorPage';

const Speakers = () => {
    const [error, setError] = useState(null);

    useEffect(() => {
        axios.get('/api/v1/speakers')
            .then(response => {
                // handle success
            })
            .catch(error => {
                setError(error.response.data.error);
            });
    }, []);
```

```
    if (error) {
        return <ErrorPage error={error} />;
    }

    return (
        // rest of your application
    );
};


export default Speakers;
```

Next, we will discuss how to track and log events in Flask web applications.

Tracking events in your application

Flask allows you to track events in your application in an elegant way. This is critical to identifying potential issues. By tracking events, you can get a better understanding of what is happening in your application and make informed decisions about how to improve the situation.

There are several ways to track events in Flask, including using built-in logging functionality, third-party logging services, or custom code tracking. For instance, you can use the Python `logging` module to log information about your application activities to a file or to the console.

Using the logging module is easy; simply import `logging` into your Flask application and configure it to log information at the appropriate level. For instance, the following code configures the logging module to log information to a file named `error.log`:

```python
import logging
from flask import Flask
app = Flask(__name__)
# Set up a logger
logger = logging.getLogger(__name__)
logger.setLevel(logging.DEBUG)

# Specify the log file
file_handler = logging.FileHandler('error.log')
file_handler.setLevel(logging.DEBUG)

# Add the handler to the logger
logger.addHandler(file_handler)

@app.route('/logger')
def logger():
    logger.debug('This is a debug message')
```

```
        logger.info('This is an info message')
        logger.warning('This is a warning message')
        logger.error('This is an error message')
        return 'Log messages have been written to the log file'

 if __name__ == '__main__':
     app.run()
```

The preceding code demonstrates the implementation of the logging module in a Flask web application. The code sets up a logger object using the `logging.getLogger(__name__)` method. The logger is set to the debug level with `logger.setLevel(logging.DEBUG)`. A `FileHandler` object is created with `file_handler = logging.FileHandler('error.log')`, and the handler is set to the debug level as well with `file_handler.setLevel(logging.DEBUG)`.

The handler is added to the logger object with `logger.addHandler(file_handler)`. In the `logger()` function, there are four logging methods called `debug()`, `info()`, `warning()`, and `error()`. These methods log messages to the log file with the respective log levels (debug, info, warning, and error). The messages logged are simple string messages.

Furthermore, when tracking events in Flask applications, you can use a third-party logging service. Using third-party logging services with Flask can provide more advanced logging features such as centralized log management, real-time log searching, and alerting.

For instance, you can use cloud-based log management services such as **AWS CloudWatch**, **Loggly**, and **Papertrail**.

Let's examine briefly the implementation of AWS CloudWatch. AWS CloudWatch is a logging service that provides log management and monitoring for AWS resources. To use AWS CloudWatch with Flask, you can use the **CloudWatch Logs** API to send log data directly to AWS CloudWatch.

The following steps implement logging in Flask applications using AWS CloudWatch:

1. Set up an AWS account and create a **CloudWatch Log Group**.
2. Install the `boto3` library, which provides a Python interface to the AWS CloudWatch API. Install `Boto2` with `pip install boto3` and ensure your virtual environment is activated.
3. In your Flask application, import the `boto3` library and configure it with your AWS credentials.
4. Create a logger and set its log level to the desired level of verbosity.
5. In your application code, use the logger to log messages at various levels such as info, warning, error, and so on.
6. Configure the logger to send logs to AWS CloudWatch. This can be done by creating a custom handler that sends log messages to CloudWatch using the `boto3` library.
7. Deploy your Flask application and monitor your logs in AWS CloudWatch.

Let's explore the code implementation:

```python
import boto3
import logging
from flask import Flask
app = Flask(__name__)
boto3.setup_default_session(
    aws_access_key_id='<your-access-key-id>',
    aws_secret_access_key='<your-secret-access-key>',
    region_name='<your-region>')
logger = logging.getLogger(__name__)
logger.setLevel(logging.DEBUG)
cloudwatch = boto3.client('logs')
log_group_name = '<your-log-group-name>'
class CloudWatchHandler(logging.Handler):
    def emit(self, record):
        log_message = self.format(record)
        cloudwatch.put_log_events(
            logGroupName=log_group_name,
            logStreamName='<your-log-stream-name>',)


if __name__ == '__main__':
    app.run()
```

The full source code can be found on GitHub.

The preceding code shows the implementation of how to use the boto3 library to send logs from a Flask application to AWS CloudWatch. It works as follows:

1. The boto3 library is imported and a default session is set up with the specified AWS access key ID, secret access key, and region name.

2. A logger object is created using the logging module and the logging level is set to DEBUG.

3. A CloudWatch client object is created using the boto3 library.

4. A custom handler class named CloudWatchHandler is created that inherits from the logging.Handler class and overrides its emit method. In the emit method, the log message is formatted and sent to AWS CloudWatch using the put_log_events method of the CloudWatch client.

5. An instance of the CloudWatchHandler class is created and its logging level is set to DEBUG. This handler is then added to the logger object.

6. A route named /logging_with_aws_cloudwatch is created that generates log messages of different levels (debug, info, warning, and error) using the logger object.

Handling errors and tracking events in your Flask application is crucial to ensuring its reliability and robustness. With Flask's built-in debugger, error handlers, custom error pages, logging, and third-party logging libraries, you can easily diagnose and resolve problems as they show up in Flask application development.

Now that you are able to implement Flask's built-in debugger, error handlers, custom error pages, logging, and third-party logging libraries, wouldn't it be nice if you had the means for the admins to receive email messages about errors in your application logs in real time?

Let's work through how this can be implemented in Flask.

Sending error emails to administrators

Sending error emails to administrators provides an efficient way to notify them about errors and issues in your Flask application. This allows you to quickly identify and resolve problems before they escalate into bigger issues and negatively impact the user experience. The benefits include timely identification and resolution of errors, improved system reliability, and reduced downtime.

Let's delve into an example of sending error emails to administrators:

```python
import smtplib
from email.mime.text import MIMEText
from flask import Flask, request

app = Flask(__name__)

def send_email(error):
    try:
        msg = MIMEText(error)
        msg['Subject'] = 'Error in Flask Application'
        msg['From'] = 'from@example.com'
        msg['To'] = 'to@example.com'

        s = smtplib.SMTP('localhost')
        s.send_message(msg)
        s.quit()
    except Exception as e:
        print(f'Error sending email: {e}')

@app.errorhandler(500)
def internal_server_error(error):
    send_email(str(error))
    return 'An error occurred and an email was sent to the
```

```
        administrator.', 500

if __name__ == '__main__':
    app.run()
```

The preceding code demonstrates the implementation of sending error emails to notify administrators about errors in a Flask application. It works as follows:

1. The code uses the `smtplib` and `email.mime.text` libraries to create and send an email message.

2. The `send_email(error)` function takes an error message as a parameter and creates an email message using the `MIMEText` object. The `subject`, `sender email address`, `recipient email address`, and `error message` are set for the email. The email is then sent using the `smtplib` library through the local email server.

Flask's `errorhandler` decorator is used to catch any `500` internal server errors that occur in the application. The `internal_server_error` function is called when an error `500` occurs, and it calls the `send_email` function with the error message passed as a parameter. The function returns a response to the user indicating that an error occurred and an email was sent to the administrator.

Summary

Error handling has been an essential aspect of software development from time immemorial. It is crucial to ensure that your Flask web application can handle errors effectively. We discussed the Flask debugger, error handlers, and custom error pages. With these, you can provide meaningful feedback to users and help maintain the stability and reliability of your application.

As full stack developers, we reinforced the importance of keeping in mind that error handling is a continuous process. You should regularly review and update your error-handling strategies to ensure that your application remains robust and resilient. We also considered logging errors and sending notifications to administrators so that you can quickly identify and resolve any issues that may arise.

In short, a bug-free development experience remains a mirage for any professional developer. You should be prepared to effectively handle expected and unexpected errors in your web applications. By doing so, your application will continue to deliver value to your users, even in the face of unexpected errors and failures.

Next, we will explore modular development in Flask using **Blueprints**. With Blueprints and modular architecture, you can easily maintain and scale your React-Flask web applications.

14
Modular Architecture – Harnessing the Power of Blueprints

In a far-off kingdom called Flaskland, there lived a brave prince named Modular. He was known for his love of clean and organized programming code, and his dream was to create a kingdom where all the snippet code lived in harmony. One day, as he was wandering through the land, he came across a castle in disarray. The code snippets were scattered everywhere, and there was no clear structure to be found.

The prince knew that this was a challenge he had to take on. He rallied his army of helper functions and organized them into modules, each with a specific purpose. He then declared that these modules were the building blocks of the kingdom, and with them, they could conquer the chaos.

And so, the prince and his army of helper functions set out to build a kingdom of well-structured, reusable code. They worked day and night until, finally, the newly organized kingdom was born. The snippets were organized, and the kingdom was a beauty to behold. This story captures the heart of code modularity, the practice of breaking down a program or system into smaller, self-contained modules or components. Blueprints in Flask encourage this modular approach to building web applications.

Modular architecture has become increasingly important as web applications have become more complex in size and scope. Modular architecture is a modular programming paradigm that emphasizes breaking down large applications into smaller, reusable modules that can be developed and tested independently.

The **object-oriented programming (OOP)** revolution of the 1980s also had a significant impact on the development of modular architecture. OOP encouraged the creation of self-contained, reusable objects that could be combined to form complex applications. This approach was well suited to the development of modular applications and helped to drive the widespread adoption of modular architecture.

The principles of modularity, separation of concerns, and encapsulation remain key elements of modular architecture, and the pattern continues to evolve and adapt to meet the changing needs of software development. Today, modular architecture is a widely accepted and widely used software design pattern.

Modular architecture is used in a variety of contexts, from large-scale enterprise applications to small, single-page web applications. In Flask web applications, Blueprints refer to a way to organize a group of related views and other code into a single module. Blueprints resemble what components are to React: reusable pieces of UI that encapsulate a set of functions and states. But in the context of Flask, Flask allows you to organize your application into smaller, reusable components called Blueprints.

In this chapter, we will explore modular architecture in web development. With Blueprints in perspective, we will discuss how Blueprints can help you to build decoupled, reusable, maintainable, and testable Flask web applications.

In this chapter, we will cover the following topics:

- Understanding the benefits of modular architecture in web development

- Understanding Flask Blueprints

- Setting up a Flask application with Blueprints

- Handling the React frontend with Flask Blueprints

Technical requirements

The complete code for this chapter is available on GitHub at: `https://github.com/PacktPublishing/Full-Stack-Flask-and-React/tree/main/Chapter14`.

Due to the page count constraints, most of the long code blocks have been snipped. Please refer to GitHub for the complete code.

Understanding the benefits of modular architecture in web development

Modular architecture is a software development approach that involves breaking down a large, complex system into smaller, independent, and reusable modules. In the history of web development, modular architecture became more apparent. The traditional monolithic architecture involved having all the components of a web application tightly coupled, resulting in a large, unwieldy code base that was difficult to maintain and scale.

As web applications became more complex and the need for scalability increased, developers began to seek alternative approaches that would allow them to break down a web application into smaller, independent components.

Modular architecture emerged as a solution to these limitations, as it allowed developers to create smaller, reusable components that could be combined to form a complete web application. This approach provided several benefits, including improved maintainability, easier scalability, and better separation of concerns.

With modular architecture, developers could work on individual components in isolation, which reduced the risk of breaking the entire application and made it easier to test and deploy changes independently. As a result, modular architecture quickly gained popularity among web developers, and many modern web development frameworks such as Flask, Django, Ruby on Rails, and Angular have embraced this architectural style. The popularity of modular architecture has continued to grow over the years, and it remains a crucial component of modern web development practices.

Let's explore some of the benefits of modular architecture in web development:

- **Scalability**: In a traditional monolithic architecture, as an application grows, it becomes more difficult to manage, maintain, and scale. With modular architecture, each module is independent and can be developed, tested, and deployed independently, which makes it easier to scale individual components as needed.

- **Reusability**: Modular architecture encourages code reuse, which leads to a more efficient development process. Modules can be reused across different projects, reducing the amount of time and effort required to develop new applications. Furthermore, modular architecture makes it easier to update and maintain existing code, as changes can be made to a single module without affecting the rest of the application.

- **Maintainability**: With modular architecture, the application is divided into smaller, manageable components, making it easier to identify and resolve issues. The modular design makes it easier to isolate problems and debug issues, reducing the time and effort required to resolve them. Furthermore, modular architecture makes it easier to test individual components, ensuring that the application remains reliable and maintainable over time.

- **Flexibility**: Modular architecture allows developers to easily modify or extend the functionality of an application without affecting the rest of the system. This makes it easier to add new features, make changes, or integrate new technologies into the application. With modular architecture, developers can work on individual modules, ensuring that the application remains flexible and adaptable over time.

- **Improved collaboration**: Modular architecture enables developers to work on different parts of an application in parallel, improving collaboration and reducing the time required to complete projects. The modular design allows teams to divide the work into smaller, manageable components, making it easier to coordinate and integrate their efforts.

- **Better performance**: Modular architecture can improve the performance of web applications by reducing the size of individual components and improving the load times of an application. With smaller, more focused components, the application can load faster, improving the user experience. Additionally, modular architecture allows for better resource allocation, ensuring that the application uses resources efficiently and effectively.

In conclusion, modular architecture is becoming increasingly important in web development, as it provides several benefits over traditional monolithic architecture. With its ability to improve scalability, reusability, maintainability, flexibility, collaboration, and performance, modular architecture provides a compelling reason for developers to adopt this approach in their projects.

By embracing modular architecture, developers can create better, more efficient applications that are easier to manage and maintain over time.

Next, we will discuss the big elephant in the Flask community – Blueprint. Blueprint is a powerful organizational tool that facilitates the structuring of a web application into modular and reusable components.

Understanding Flask Blueprints

As you may be aware, Flask is a simple and lightweight framework that allows developers to create web applications quickly and easily. Flask Blueprints are an important feature of Flask that help developers organize their applications into reusable components.

Flask Blueprints are a way to organize your Flask application into smaller and reusable components. Essentially, Blueprints are a collection of routes, templates, and static files that can be registered and used in multiple Flask applications. Blueprints allow you to split your Flask application into smaller, modular components that can be easily maintained and scaled. This modular approach to building web applications makes it easier to manage the code base and collaborate with other developers.

Let's glance through some of the benefits of using Blueprints in your Flask application development:

- **Modular design**: Flask Blueprints allow developers to break down their applications into smaller, reusable components. This makes it easier to maintain and scale the code base over time.

- **Reusability**: Once you create a Blueprint, you can reuse it across different Flask applications. This saves you time and effort. Indeed, using Flask Blueprints can greatly simplify the process of building complex web applications, allowing developers to quickly and easily create reusable components with just a few clicks of the mouse.

- **Flexibility**: Flask Blueprints can be customized to suit the needs of your application. You can define your own URL prefixes for a Blueprint, which allows you to customize your application's URL structure. This gives you more control over how your web application is structured and accessed.

- **Template inheritance**: Blueprints can inherit templates from the main application, which allows you to reuse templates across multiple Blueprints. This makes it easier to create consistent and well-designed web applications.

- **Namespaces**: Blueprints can define their own view functions, and these functions are namespaced within the Blueprint. This helps prevent naming conflicts between different parts of your application.

Flask Blueprints undoubtedly promote a clear separation of concerns within your application code base. By organizing your code into separate Blueprints, you can ensure that each component of your application is responsible for a specific area of functionality. This can make it easier to understand and debug your code, as well as ensure that your application is more maintainable over time.

In the next section, we will delve into setting up a Flask application with Blueprints in mind.

Setting up a Flask application with Blueprints

Blueprints in Flask are a way to organize and structure a Flask application into smaller, reusable components. To use Blueprints in a Flask application, you typically define your Blueprint in a separate Python file where you can define your routes, templates, and any other necessary logic specific to that Blueprint. Once defined, you can register the Blueprint with your Flask application, which allows you to use the Blueprint functionality within your main Flask application.

With Blueprints, you can easily separate concerns between different parts of your application, making it easier to maintain and update over time.

Now, let's dive deep into the heart of how you can set up Flask applications with Blueprints.

Structuring Blueprint Flask applications

In web application development, efficient organization and modularity of the code base are essential for building robust and maintainable projects. One of the key structural elements in Flask is the concept of Blueprints. These Blueprints provide a structured way to compartmentalize and encapsulate various components of a web application.

The approach invariably promotes clarity, reusability, and scalability. We are going to examine the structure of the `attendees` Blueprint – a crafted, organizational structure designed to streamline the development of attendee-related features within our web application.

The `attendees` Blueprint is nestled within the `bizza\backend\blueprints\attendees` directory. Create a new directory inside the `bizza/backend` project directory for the Flask applications and name it `blueprints`. The Blueprints added to the project make the directory structure appear as follows:

The attendees Blueprint:

```
bizza\backend\blueprints\attendees
-models
-templates
-static
-attendee_blueprint.py
```

Detailed structure:

```
bizza\backend\blueprints\attendees
-models
- __init__.py
- attendee.py
-templates
- attendees/
- base.html
- attendee_form.html
- attendee_list.html
- attendee_profile.html
- attendee_profile_edit.html
-static
- css/
- attendees.css
- js/
- attendees.js
attendee_blueprint.py
```

The preceding `attendees` Blueprint contains the following components:

- `models`: This is a subdirectory containing a Python module named `attendee.py` that defines the data model for attendees. The `__init__.py` file is a blank Python module that indicates to Python that this directory should be treated as a package.

- `Templates`: This is a subdirectory containing HTML templates for the attendee views. The `base.html` template is a base template that other templates inherit from. The `attendee_form.html` template is used for creating or editing attendee profiles. The `attendee_list.html` template is used to display a list of all attendees. The `attendee_profile.html` template is used to display a single attendee's profile. The `attendee_profile_edit.html` template is used to edit an attendee's profile.

- `static`: This is a subdirectory containing static files used by the templates. The `css` directory contains an `attendees.css` file used to style the HTML templates. The `js` directory contains an `attendees.js` file used for client-side scripting.

- `attendee_blueprint.py`: This is a Python module containing the Blueprint definition and the routing for the attendee views. This Blueprint defines routes for displaying a list of attendees, displaying an individual attendee's profile, creating a new attendee profile, and updating an existing attendee profile. The Blueprint also contains database-related functions for handling attendee data, such as adding new attendees and updating attendee information.

Defining models and Blueprint modules

Models serve as the foundation of a web application's data structure. Models represent the essential entities and relationships within web applications. They encapsulate data attributes, business logic, and interactions, providing a coherent representation of real-world concepts.

When defining models within a Blueprint module, you create a self-contained unit that encapsulates data-related logic. With the integration of models into Blueprint modules, you achieve a harmonious synergy, and benefits such as the following:

- **Clear separation**: Blueprint modules isolate various functionalities, while models encapsulate data handling. This separation simplifies code base maintenance and enhances readability.

- **Coherent structure**: Blueprint modules provide a logical context for models, making it easier to navigate and understand data-related operations.

- **Reusability**: Models defined within a Blueprint can be reused across other parts of the application through Blueprint integration, promoting a **Don't Repeat Yourself (DRY)** coding approach.

Now, let's delve into the attributes of the attendee model in a Blueprint module:

Attendees Blueprint:

```
-models
-  __init__.py
-  attendee.py
```

The `attendee.py` model is defined as the following:

```python
from bizza.backend.blueprints import db


class Attendee(db.Model):
    id = db.Column(db.Integer, primary_key=True)
    name = db.Column(db.String(50), nullable=False)
    email = db.Column(db.String(120), unique=True,
        nullable=False)
    registration_date = db.Column(db.DateTime,
        nullable=False)

    def __repr__(self):
        return f'<Attendee {self.name}>'
```

The preceding `Attendee` model represents attendees at the conference. It has columns for `id`, `name`, `email`, and `registration_date`. The `__repr__` method specifies how instances of the model should be represented as strings.

The attendee Blueprint is defined as follows:

```
from bizza.backend.blueprints.attendees.models.attendee import
Attendee
from bizza.backend.blueprints.attendees.forms import AttendeeForm,
EditAttendeeForm
from bizza.backend.blueprints.attendees import db
attendee_bp = Blueprint('attendee', __name__, template_
folder='templates', static_folder='static')

@attendee_bp.route('/attendees')
def attendees():
    attendees = Attendee.query.all()
    return render_template('attendees/attendee_list.html',
        attendees=attendees)
@attendee_bp.route('/attendee/add', methods=['GET',
    'POST'])

def add_attendee():
    form = AttendeeForm()
    if form.validate_on_submit():
        attendee = Attendee(name=form.name.data,
                            email=form.email.data,
                            phone=form.phone.data,
            ...
        return redirect(url_for('attendee.attendees'))
    return render_template('attendees/attendee_form.html',
        form=form, action='Add')
...
```

The preceding snippet defines a Flask Blueprint for managing attendees. It imports the necessary modules, including the `Attendee` model, `AttendeeForm`, and `EditAttendeeForm` from the `attendees` package, and db from the `bizza.backend.blueprints` package.

The Blueprint has a route for the attendee list that requires the user to be logged in. It retrieves all attendees from the database using the `Attendee.query.all()` method and renders the `attendee_list.html` template with the list of attendees.

The Blueprint also has a route for adding attendees that is accessible via GET and POST requests. It creates an instance of `AttendeeForm`, and if the form is validated, it creates a new attendee object with the data submitted through the form, adds it to the database, and commits the changes. If successful, it flashes a message and redirects to the attendee list page. If the form is not valid, it re-renders the `attendee_form.html` template with the form and the *Add* action.

Registering the Blueprints

When you create a Blueprint, you define its routes, views, models, templates, and static files. Once you have defined your Blueprint, you need to register it with your Flask application using the `register_blueprint` method. This method tells Flask to include the views, templates, and static files of the Blueprint in the application.

So, when the `app.register_blueprint` method is called, it adds the routes and views defined in the Blueprint to the application. This makes the functionality provided by the Blueprint available to the rest of the application.

Let's use a basic Flask application factory function to create and configure a Flask application:

```python
from flask import Flask
from flask_sqlalchemy import SQLAlchemy

# initialize the db object
db = SQLAlchemy()

def create_app():
    app = Flask(__name__)

    # load the config
    app.config.from_object('config.Config')

    # initialize the db
    db.init_app(app)

    # import the blueprints
    from .blueprints.speaker_blueprint import speaker_bp
    from .blueprints.presentation_blueprint import
        presentation_bp
    from .blueprints.attendee_blueprint import attendee_bp


    # register the blueprints
    app.register_blueprint(speaker_bp)
    app.register_blueprint(presentation_bp)
    app.register_blueprint(attendee_bp)

    return app
```

The preceding code does the following:

1. Imports the `Flask` and `SQLAlchemy` modules.

2. Creates an instance of the Flask application.

3. Loads the configuration from a configuration file.

4. Initializes the `SQLAlchemy` object with the application.

5. Imports the Blueprints from the different parts of the application.

6. Registers the Blueprints with the Flask application.

7. Returns the Flask application object.

Next, we will shift our focus to how Blueprints and the React frontend can be integrated seamlessly. We need to get creative and discover exciting ways to blend Blueprints with a React frontend and take our development to the next level.

Handling the React frontend with Flask Blueprints

In the case of a React frontend and Flask backend, Blueprints can be used to organize the different API routes and views that the frontend needs to communicate with the backend. The frontend can make requests to the backend API endpoints that are defined in the Blueprints, and the backend can respond with the appropriate data.

Additionally, using Flask for the backend and React for the frontend provides a flexible and powerful development environment. Flask is a lightweight and *easy-to-use* web framework that is ideal for building **RESTful** APIs, while React is a popular and powerful frontend library that allows for the creation of complex and dynamic user interfaces. With these technologies, you can create high-performance, scalable web applications that are easy to maintain and update.

It's time to unleash our imagination and explore the limitless potential of combining a Blueprint with a React frontend. Integrating a Flask backend with a React frontend involves setting up the communication between the two using API endpoints. We set up a typical Flask Blueprint, for instance, the `attendees` Blueprint structure, as follows:

```
bizza\backend\blueprints\attendees
-models
-attendee_blueprint.py
```

This route should serve as the entry point to the React app. Modify the existing Flask routes in `attendees_blueprint.py` to return JSON data instead of HTML.

In the React frontend, we will create an `attendee` component and make API calls to the Flask routes using a library such as `axios` to retrieve the JSON data and render it in the UI.

The updated `attendee_blueprint.py` file is as follows:

```
from flask import Blueprint, jsonify, request
from bizza.backend.blueprints.attendees.models.attendee import
Attendee
from bizza.backend.blueprints.attendees.forms import AttendeeForm,
EditAttendeeForm
from bizza.backend.blueprints.attendees import db


attendee_bp = Blueprint('attendee', __name__, url_prefix='/api/
attendees')



@attendee_bp.route('/', methods=['GET'])
def get_attendees():
    attendees = Attendee.query.all()
    return jsonify([a.to_dict() for a in attendees])



@attendee_bp.route('/<int:attendee_id>',
    methods=['DELETE'])
def delete_attendee(attendee_id):
    attendee = Attendee.query.get_or_404(attendee_id)
    db.session.delete(attendee)
    db.session.commit()
    return jsonify(success=True)
```

The preceding code defines a Flask Blueprint for managing attendees in the application. The Blueprint is registered at the `/api/v1/attendees` URL prefix. It includes routes for getting all attendees, adding a new attendee, getting a specific attendee, updating an existing attendee, and deleting an attendee.

The `get_attendees()` function is decorated with `@attendee_bp.route('/', methods=['GET'])`, which means it will handle GET requests to the `/api/v1/attendees/` URL. It queries the database for all attendees, converts them into a dictionary using the `to_dict()` method defined in the `Attendee` model, and returns a JSON representation of the list of attendees.

The `add_attendee()` function is decorated with `@attendee_bp.route('/', methods=['POST'])`, which means it will handle POST requests to the `/api/v1/attendees/` URL. It first creates an `AttendeeForm` object from the POST request data. If the form data is valid, a new attendee is created using the form data and added to the database.

The new attendee is then converted into a dictionary using the `to_dict()` method and returned as a JSON response. If the form data is not valid, the errors are returned as a JSON response.

The `get_attendee()` function is decorated with `@attendee_bp.route('/<int:attendee_id>', methods=['GET'])`, which means it will handle `GET` requests to the `/api/v1/attendees/<attendee_id>` URL, where `attendee_id` is the ID of the specific attendee being requested. It queries the database for the attendee with the specified ID, converts it into a dictionary using the `to_dict()` method, and returns a JSON representation of the attendee.

The `update_attendee()` function is decorated with `@attendee_bp.route('/<int:attendee_id>', methods=['PUT'])`, which means it will handle `PUT` requests to the `/api/v1/attendees/<attendee_id>` URL. It first queries the database for the attendee with the specified ID. It then creates an `EditAttendeeForm` object from the `PUT` request data, using the current attendee object as the default value.

If the form data is valid, the attendee object is updated with the new data and saved to the database. The updated attendee object is then converted into a dictionary using the `to_dict()` method and returned as a JSON response. If the form data is not valid, the errors are returned as a JSON response.

The `delete_attendee()` function is decorated with `@attendee_bp.route('/<int:attendee_id>', methods=['DELETE'])`, which means it will handle `DELETE` requests to the `/api/v1/attendees/<attendee_id>` URL. It queries the database for the attendee with the specified ID, deletes it from the database, and returns a JSON response indicating success.

The utilization of Flask Blueprints to handle the integration of a React frontend with a Flask backend offers numerous benefits in terms of code organization, modularity, scalability, and maintainability. This structured development approach facilitates seamless full stack development while maintaining a clear separation of concerns.

Summary

As we come to the end of this chapter, let's take a moment to reflect on the exciting journey we've been on. The chapter explores modular architecture in web development and how Flask Blueprints can help build decoupled, reusable, maintainable, and testable Flask web applications.

The benefits of modularity, separation of concerns, and encapsulation remain key elements of modular architecture. In Flask, Blueprints organize a group of related views and other code into a single module. This chapter also covers setting up a Flask application with Blueprints. Finally, we discussed a very flexible way to build full stack web applications at scale with React frontend and Flask Blueprints.

Next, we will explore unit testing in Flask. Fasten up and let's delve into the exciting world of testing in Flask backend development.

15

Flask Unit Testing

Unit testing is an essential phase in software development that guarantees the proper functioning of each component of an application. In *Chapter 7, React Unit Testing*, we discussed unit testing as it relates to React components in building reliable user interfaces for the frontend part of a web application. With backend development, the principles of unit testing are similar, except that you are using a different programming language – or better, still working with a backend tech stack.

Unit testing ensures that each component or module of a software application is working correctly in isolation from the rest of the application. By testing each unit separately and thoroughly, developers can identify and fix issues early in the development cycle, which can save time and effort in the long run.

Unit testing helps catch defects early and provides a safety net for refactoring code, making it easier to maintain and evolve the application over time. Ultimately, the goal of unit testing is to produce high-quality software that meets the requirements and expectations of end users.

In this chapter, we will discuss briefly the importance of unit testing in Flask and explore the benefits of using pytest as a testing framework for Flask applications. We will also cover the installation and setup process for pytest, as well as the fundamentals of **test-driven development** (TDD).

Additionally, we will delve into writing basic tests and assertions and handling exceptions. At the end of this chapter, you will be able to understand the importance of unit testing in Flask applications, describe what pytest is and how it differs from other testing frameworks, and how pytest can be integrated into your existing project.

You will have also learned how to test JSON APIs using pytest and understand how to make requests to the API endpoints and validate the response data. Finally, you will be able to apply TDD principles to write tests before writing the actual code and use the tests to guide the development process.

In this chapter, we'll be covering the following topics:

- Unit testing in Flask applications
- Introducing pytest
- Setting up of pytest

- Basic syntax, structures, and features of pytest
- Writing unit tests
- Testing JSON APIs
- Test-driven development with Flask
- Handling exceptions

Technical requirements

The complete code for this chapter is available on GitHub at: `https://github.com/PacktPublishing/Full-Stack-Flask-and-React/tree/main/Chapter15`

Unit testing in Flask applications

Flask is like a chef's knife for web developers – it's a versatile tool that can help you cook up scalable and flexible applications in no time. However, as the complexity of Flask applications grows, it becomes increasingly difficult to ensure that all the components of the application are working correctly together. This is where unit testing comes in.

Unit testing is a software testing technique that involves testing each component or module of an application in isolation from the rest of the application. By testing each unit separately and thoroughly, developers can identify and fix issues at the outset of the development process. The practice of unit testing can assist in spotting defects quickly and serve as a safeguard when making changes or modifying code, thus making it easier to maintain and evolve the application over time.

With Flask applications, unit testing helps ensure that all the routes, views, and other components are working as expected. Unit testing can also help catch issues with database interactions, external API calls, and other external dependencies.

The testing heuristics or principles are as follows:

- **FIRST**: Fast, Independent, Repeatable, Self-Validating, and Timely
- **RITE**: Readable, Isolated, Thorough, and Explicit
- **3A**: Arrange, Act, Assert

These principles can be utilized by developers as guidelines and best practices to ensure the effectiveness of their unit testing efforts. These testing principles can enhance the quality of code, minimize bugs and defects, and ultimately deliver superior software products to application users. By adhering to these principles, developers and testers can improve the overall reliability and maintainability of the code base.

Let's briefly examine these testing principles to understand how they can guide you in writing excellent unit tests.

FIRST

FIRST emphasizes the importance of unit tests being quick to run, not dependent on external factors, able to be run repeatedly without side effects, self-checking, and written promptly:

- **Fast**: Unit tests should be designed to execute quickly so that they can be run frequently without delaying the development cycle. This means that the unit under test should be lightweight and should not depend on external resources, such as databases or network connections, which can introduce delays in the test execution. In Flask, we can ensure fast execution of tests by mocking external dependencies using tools such as the `pytest_mock` plugin.

- **Independent**: Unit tests should be designed to run independently of each other so that the failure of one test does not affect the execution of other tests. In Flask, we can achieve independence between tests by resetting the application state before each test using the Flask test client.

- **Repeatable**: Unit tests should be designed to produce the same result every time they are run, regardless of the environment in which they are executed. This means that the unit under test should not rely on external factors, such as system time or random number generators, that can introduce variability in the test results.

- **Self-checking**: Unit tests should be designed to check their results and report failures without requiring human intervention. This means that the unit test should include assertions that compare the expected results with the actual results of the test. In Flask, we can use the built-in assert statement to check the test results.

- **Timely**: Unit tests should be designed to be written promptly, ideally before the code they are testing is written. This means that they should be part of the development process and not an afterthought. In Flask, we can follow the TDD approach to ensure that tests are written before the code.

Next, we will explore RITE (Reproducible, Isolated, Thorough and Extensible), a testing principle that can enhance the effectiveness of unit tests and enhance code quality.

RITE

RITE emphasizes the importance of unit tests being easy to read and understand, isolated from other components, covering all possible scenarios, and explicit in their assertions:

- **Reproducible**: Tests should be able to be reproduced on different systems and environments. This means that tests should not rely on external factors such as network connectivity, time, or other system resources. By ensuring that tests can be run consistently across different environments, developers can be confident that their code works as intended.

- **Isolated**: Tests should be independent of each other and not share any state. This means that each test should start with a clean slate and not rely on any previous test results or global state. By isolating tests, developers can ensure that each test is testing a specific piece of functionality and is not affected by other parts of the system.

- **Thorough**: Tests should test all aspects of the system, including edge cases and error conditions. This means that developers should strive to create tests that cover as much of the code base as possible, including all possible inputs and outputs.

- **Extensible**: Tests should be easy to extend and maintain as the system evolves. This means that tests should be designed to accommodate changes in the code base, such as new features or changes in the system architecture.

In a nutshell, the RITE principles are beneficial because they can help you to improve the quality, reliability, and maintainability of your code.

Moving forward, we will explore 3A (Arrange, Act, and Assert), a unit test approach that can make your unit tests more readable and maintainable.

3A

3A is a simple guideline for structuring a unit test and consists of three steps – Arrange, Act, and Assert. The Arrange phase sets up the test scenario, the Act phase performs the action being tested, and the Assert phase checks the expected outcome. The 3A principle is the best practice for designing and writing effective unit tests:

- **Arrange**: In this step, you set up the conditions for the test by initializing objects, setting variables, and other necessary actions. This ensures that the test environment is properly configured and that the system under test is in the expected state.

- **Act**: In this step, you perform the action or method call that is being tested. This may involve passing arguments to a function, invoking a method on an object, or making a request to an API endpoint. The key is to ensure that the action being taken is specific and targeted at the functionality being tested.

- **Assert**: In this step, you verify that the outcome of the action matches the expected result. This often involves checking the value returned by a function, comparing the state of an object before and after a method call, or ensuring that an API endpoint returns the correct response status code and data.

Next, we will explore Pytest as a widely used testing framework that seamlessly integrates with Flask. Pytest is empowering developers to efficiently create and execute unit tests, integration tests, and more, ensuring the robustness and reliability of Flask web applications.

Introducing Pytest

Pytest is an open source testing framework for Python that simplifies the process of writing and executing concise and readable tests. Pytest provides a simple and flexible way to write tests and supports a wide range of testing options out of the box, including functional tests, unit tests, and integration tests.

Pytest is widely used among Python developers due to its ease of use, powerful fixture system, and integration with other Python testing tools. Pytest can automatically find and run all the tests in a project with the -test discovery ability. Pytest generates detailed reports that provide developers with valuable insights into the test results.

These reports include information on the number of tests executed, the time taken to run each test, and any failures or errors that occurred. This information can help developers pinpoint and address issues promptly, improving the overall quality of the code base. Pytest has an amazing large community of users and contributors who actively develop and maintain plugins that extend Pytest functionalities.

Interestingly, Pytest differs from other testing frameworks such as `unittest`, `nose`, `doctest`, `tox`, `hypothesis library`, and `robot framework` with its simplicity and power, versatility, and community support, providing easy-to-use testing capabilities with detailed reporting. Pytest is undoubtedly a popular choice among Python developers for unit testing and other testing needs.

Next, we'll walk through the steps of setting up Pytest and creating our first test.

Setting up Pytest

Testing your Python code is an essential part of the development process, and Pytest is a powerful tool for actualizing a robust testing environment. In this section, we'll walk you through the steps of setting up Pytest and transforming your Python code testing experience from amateur into pro, providing advanced features and capabilities that make testing faster, easier, and more effective.

To set up Pytest, you can follow these steps:

1. **Installing pip**: You can install Pytest using `pip`, the package installer for Python. Open your Terminal or command prompt in the `bizza/backend/` project directory and run the following command:

    ```
    pip install pytest
    ```

    The preceding line installs Pytest and all its dependencies.

2. **Creating a test file**: Create a new file named `test_addition.py` in your project directory – that is, `bizza/backend/tests/test_addition.py`. This is a simple example test file to warm up with.

3. **Writing a test function**: Inside `test_addition.py`, write a simple test function using the following format:

    ```
    def test_function_name():
        assert expression
    ```

Let's discuss the preceding short format snippet:

- `test_function_name` represents the test function's name
- `expression` represents the code you want to test
- The `assert` statement checks whether the expression is true and raises an error if the expression is false

> **Note**
>
> In Pytest, test functions are identified by their name and should start with the `test_` prefix. With this naming convention, Pytest can recognize your functions as tests and run them automatically. When you run Pytest in the Terminal, Pytest searches your code base for any functions that begin `test_`. Then, Pytest executes those functions and reports the results of the tests.

Now, let's describe a test function that tests whether adding two numbers produces the expected result:

```
def test_addition():
    assert 1 + 1 == 2
```

The preceding code shows a simple Pytest test function that tests the addition of two numbers. The function's name starts with `test_`, which tells Pytest that it is a test function.

The body of the function contains an assertion that checks whether $1 + 1$ equals 2. If the assertion is `true`, then the test passes. If the assertion is `false`, then the test fails and Pytest reports an error.

4. **Running the tests**: Open your Terminal and navigate to your project directory – that is, `bizza/backend/`. Run the following command to run your tests:

    pytest

    The `pytest` command will automatically discover and run all the tests in your project. You should see a message indicating that your test passed. Hats off – you have set up Pytest and run your first test! You can continue to add more test functions to your `test_addition.py` file using the same format.

    The following Terminal output shows the test result as passed:

```
(venv) C:\bizza\backend>pytest
========================================================================
========= test session starts ==========================================
========================================
platform win32 -- Python 3.10.1, pytest-7.3.1, pluggy-1.0.0
rootdir: C:\bizza\backend
plugins: Faker-16.6.0
collected 1 item
```

```
tests\test_addition.py [100%]
================================================================
========= 1 passed in 21.61s =====================================
=====================================
```

Let's take a look at the preceding output:

I.   The first line in the preceding code shows some information about the platform and versions of Python, Pytest, and other related plugins.

II.   The second line indicates the root directory for the tests. In this case, it is `C:\bizza\backend`.

III.   The third line shows that Pytest has collected one test item, which is stored in the `tests\test_addition.py` file.

IV.   The fourth line shows the result of the test: a single dot indicates that the test passed. If the test had failed, this would have been indicated by `"F"`.

V.   The fifth line shows some summary information, including the number of tests that passed, and the time taken to run the tests.

VI.   Finally, the command prompt returns, indicating that the test has finished running.

Let's assume the `test_addition.py` function's output has changed to 5 instead of 2. Should we expect the test to fail? Of course, yes! The test should fail. The following is the output of the failed test:

```
(venv) C:\bizza\backend>pytest
=================================================== test session starts
=================================================
collected 1 item
tests\test_addition.py F                                    [100%]

=================================================== FAILURES ======
=================================================
_____ test_addition


    def test_addition():
>       assert 1 + 1 == 5
E       assert (1 + 1) == 5

tests\test_addition.py:3: AssertionError
```

The preceding output indicates that the test named `test_addition.py` has failed. The assertion asserts `1 + 1 == 5` is failing because the actual result of 1 + 1 is 2, not 5.

Ready for the next step? Let's examine the basic syntax and structure of Pytest. Then, we will dive deeper into unit testing with Pytest.

Basic syntax, structures, and features of Pytest

The basic syntax and structure of a Pytest test function can be represented as follows:

```
def test_function_name():
    # Arrange: set up the necessary test data or
      environment
    # Act: execute the code being tested
    result = some_function()
    # Assert: check that the expected behavior is observed
    assert result == expected_result
```

`test_function_name` should be a descriptive name that conveys the purpose of the test:

- The `Arrange` section sets up the necessary test data or environment, such as initializing objects or connecting to a database

- The `Act` section executes the code being tested, such as calling a function or performing a specific action

- The `Assert` section checks that the expected behavior is observed, using assertions to verify that the output or behavior of the code matches what was expected

Pytest supports a wide range of assertions, including `assert x == y`, `assert x != y`, `assert x in y`, and many more. Pytest also supports the use of fixtures, which can be used to manage test dependencies and set up test data and environments.

The basic syntax and structure of a Pytest test function are designed to make it easy to write clear, concise tests that verify that your code works as expected. With Pytest's structure and the use of fixtures, you can write tests that are reliable, repeatable, and easy to maintain.

Next, we will look at one of the key Pytest features: **fixtures**.

Using fixtures

In software testing, a **fixture** is a defined state or set of data that is needed for a test to run. Essentially, fixtures are functions that help in managing and providing consistent resources, such as data, configuration, or objects, to different test cases within a test suite. Fixtures enable you to establish a stable and controlled environment for testing.

They ensure that each test case has access to the required resources without duplicating setup and teardown methods across multiple tests. You are probably wondering what setup and teardown methods are. Let's pause for a minute and shed more light on this duo in testing Flask applications.

In unit testing, the concepts of setup and teardown methods are pivotal techniques that are used to prepare and clean up the testing environment before and after the execution of each test case. Before delving into test cases, the setup procedure comes into play. The setup method is executed before each test case, and its purpose is to establish the required conditions for testing.

For instance, let's consider a Flask unit test scenario; the setup method could be designed to mimic a Flask application instance and configure a testing client, thereby providing the necessary infrastructure to simulate HTTP requests and responses for testing purposes.

On the flip side, there is the teardown phase. The teardown procedure takes place post-execution of every test case and involves cleaning up resources that were initially established during the setup operation. Back to the Flask unit test illustration, the teardown method might be programmed to gracefully terminate the testing client and shut down the Flask application instance. This ensures that no lingering resources remain active that can disrupt subsequent tests.

This duo of setup and teardown is typically located within the confines of a class encapsulating the suite of test cases. To understand it better, consider the following code snippet, which illustrates a class incorporating setup and teardown methods to validate a Flask application:

```
class FlaskTestCase:

    def setup(self):
        self.app = create_app()
        self.client = app.test_client()

    def teardown(self):
        self.app = None
        self.client = None

    def test_index_page(self):
        response = self.client.get("/")
        assert response.status_code == 200
        assert response.content == b"Bizza Web Application"
```

In the preceding code, the setup method creates a Flask application instance and a test client. On the other hand, the teardown method gracefully concludes the test client and disposes of the Flask application instance. The outcome is a neat and orderly closure of resources once a test concludes.

However, in pytest, the setup and teardown paradigms can be emulated using fixtures. Fixtures serve as functions designated to furnish shared resources to multiple test cases. Fixtures allow you to define and manage test dependencies. This is how fixtures work in pytest. You define a fixture with the @ pytest.fixture decorator. This function can then be used as a parameter in test functions, which allows the test function to access the fixture's data or environment.

When a test function is run, pytest automatically detects any fixtures that are defined as parameters and runs those fixture functions first, passing their return values as arguments to the test function. This ensures that the test function has access to the data or environment it needs to run correctly.

The following code snippet showcases a fixture that can be used to produce a Flask application instance and a test client:

```python
import pytest

@pytest.fixture()
def app():
    app = create_app()
    return app

@pytest.fixture()
def client(app):
    client = app.test_client()
    return client
```

The preceding code shows that the app fixture creates a Flask application instance and the client fixture creates a test client. These fixtures can then be used by test cases within the test suite to get access to the Flask application and the test client.

It is noteworthy to say one clear advantage of adopting fixtures for setup and teardown is their potential for reusability. By using fixtures, the setup and teardown logic can be efficiently shared across multiple test cases. This will invariably ensure that the testing code is more maintainable, and by extension, enhance the reusability of test cases.

Fixtures in your tests can provide clear benefits, including the following:

- **Reusability**: You can define a fixture once and use it in multiple tests. This can save time and reduce duplication.

- **Readability**: By separating the setup code into a fixture function, your test functions can be more focused and easier to read.

- **Maintainability**: Fixtures ensure that your tests are consistent and repeatable, even as your code base evolves.

Fixtures in pytest provide a powerful and flexible mechanism for managing test dependencies and simplifying your testing workflow.

Now, let's delve into parameterizing in pytest. Using parameterized tests in pytest allows you to test your code more thoroughly with less code duplication.

Parameterizing in pytest

Parameterizing tests in pytest is a feature that enables you to write a single test function that can be executed with different sets of input parameters. This is useful when you want to test a function or method with a variety of inputs or configurations.

To parameterize a test function in pytest, you can use the @pytest.mark.parametrize decorator. This decorator takes two arguments: the name of the parameter and a list of values or tuples representing the different parameter sets to test.

Let's explore a parameterized test function in pytest:

```
import pytest

def add(a, b):
    return a + b

@pytest.mark.parametrize("a, b, expected_result", [
    (1, 2, 3),
    (10, 20, 30),
    (0, 0, 0),
    (-1, 1, 0), ids=["1+2=3", "10+20=30", "0+0=0",
        "-1+1=0"]
])
def test_addition(a, b, expected_result):
    assert add(a, b) == expected_result
```

The preceding code is a demonstration of parameterized tests in pytest to test a function with multiple input values.

The function being tested is add(a, b), which takes two arguments, a and b, and returns their sum. The @pytest.mark.parametrize decorator is used to provide a list of input values and their corresponding expected results.

The decorator takes three arguments:

- A comma-separated string of parameter names – in this case, "a, b, expected_result".
- A list of tuples representing the parameter sets and their expected results. In this example, we have four parameter sets: (1, 2, 3), (10, 20, 30), (0, 0, 0), and (-1, 1, 0).
- An optional ids argument, which provides custom names for the test cases.

For each parameter set in the list, pytest will execute the test_addition() function with the corresponding a, b, and expected_result values. The assert statement in the test function checks that the actual result of add(a, b) matches the expected result.

When the test function is executed, pytest will generate a separate report for each parameter set, so you can see exactly which cases passed and which ones failed:

- The first parameter set, `(1, 2, 3)`, tests whether the `add()` function correctly adds 1 and 2, resulting in 3

- The second parameter set, `(10, 20, 30)`, tests whether `add()` correctly adds 10 and 20, resulting in 30

- The third parameter set, `(0, 0, 0)`, tests whether `add()` correctly adds two zeros, resulting in 0

- The fourth parameter set, `(-1, 1, 0)`, tests whether `add()` correctly adds -1 and 1, resulting in 0

Parameterizing tests can help you write more concise and effective test code by reducing the amount of duplication in your test functions and making it easier to test a wide range of inputs and configurations.

And that's not all in terms of pytest's features. Next, we'll explore mocking external dependencies in pytest.

Mocking external dependencies in pytest

Mocking external dependencies is a testing technique that involves creating simulated versions of external dependencies, such as APIs or databases, to isolate your code under test from these dependencies. When you're writing unit tests, you typically want to test only the code within the scope of the test, not any external services or libraries that it relies on.

This practice helps you keep your tests focused and fast, as well as avoid false positives or false negatives that can result from relying on external dependencies that may not be available or may behave unpredictably.

To create a mock object, you must use a mocking framework, such as `unittest.mock` or `pytest-mock`, to create a fake object that mimics the behavior of the real object. You can then use this mocked object in your tests instead of the real object, which allows you to test your code in a controlled environment.

For instance, let's say you are testing a function that retrieves data from an external API. You can use a mocking framework to create a mock object that mimics the behavior of the API, and then use this mocked object in your tests instead of making actual API calls. This allows you to test your function's behavior in a controlled environment, without you having to worry about network connectivity or the behavior of the external API.

Using a mocking strategy in your tests can also help you write more comprehensive tests as it allows you to simulate error conditions or edge cases that might be difficult or impossible to replicate with a real external dependency. For example, you can use a mocked object to simulate a network timeout or a database error, and then verify that your code under test handles these conditions correctly.

Let's say we have a `Speaker` class in our project that depends on an external `email_service` module to send email notifications to speakers. We want to write a test for the `Speaker` class that verifies that the `Speaker` class sends the expected email notifications when a new speaker is added. To achieve this, we can use the `pytest-mock` plugin to mock the `email_service` module and check that the expected calls are made.

Let's dive into a code implementation.

In the `bizza/backend/tests` directory, add the `test_speaker.py` file:

```python
# test_speaker.py
from bizza.backend.speaker import Speaker

def test_speaker_notification(mocker):
    # Arrange
    email_mock = mocker.patch(
        "bizza.backend.email_service.send_email")
    speaker = Speaker("John Darwin", "john@example.com")

    # Act
    speaker.register()

    # Assert
    email_mock.assert_called_once_with(
        "john@example.com",
        "Thank you for registering as a speaker",
        "Hello John, \n\nThank you for registering as a
        speaker. We look forward to your talk!\n\nBest
        regards,\nThe Conference Team"
    )
```

In the preceding code, we created a mocked object for the `email_service.send_email` function using `mocker.patch`. Then, we created a new `Speaker` object and called the `Speaker` object's `register()` method, which should trigger an email notification to be sent.

Then, we used the `assert_called_once_with` method of the mocked object to check that the expected email was sent with the correct arguments. If the `send_email` function is called with different arguments, the test will fail.

By using `pytest-mock` to mock the external dependency, we can isolate our test from any potential network issues or other dependencies of the `email_service` module. This makes our test more reliable and easier to maintain over time.

Mocking external dependencies is a powerful technique for isolating your code under test from external services or libraries, and for creating controlled environments that allow you to write comprehensive, reliable tests.

Writing unit tests

Writing tests with pytest involves creating test functions that verify the functionality of your code. These test functions are executed by pytest and can be organized into test modules and test packages. In addition to test functions, pytest provides other testing features such as fixtures, parameterization, and mocking, which can help you write more robust and efficient tests.

In this section, we will cover the basics of writing tests with pytest, including creating test functions, using assertions to check for expected behavior, and organizing tests into test suites.

Now, let's laser-focus on writing unit tests for a user registration component of an application.

Unit-testing user registration

Unit testing is a crucial part of the software development process. Unit testing unarguably allows developers to verify that their code works correctly and reliably, as stated earlier. One area where unit testing is particularly important is user registration, which is a critical part of many applications.

A user registration feature typically involves collecting user input, validating the input, storing it in a database, and sending a confirmation email to the user. Testing these features thoroughly is important to ensure that it works as intended and that users can register successfully and securely.

In this context, unit tests can be used to verify that the registration feature handles various scenarios correctly, such as valid and invalid inputs, duplicate usernames, and email confirmation.

Let's examine a unit test implementation for user registration.

User creation unit test

Let's test that new users can be created and saved to the database. In the `tests` directory, create `test_user_login_creation.py`:

```
def test_create_user(db):
    # Create a new user

    user = User(username='testuser',
        password='testpassword',
            email='test@example.com')
    #Add the user to the database
    db.session.add(user)
    db.session.commit()
```

```
# Retrieve the user from the database
retrieved_user = db.session.query(User)
    .filter_by(username='testuser').first()
# Assert that the retrieved user matches the original
  user
assert retrieved_user is not None
assert retrieved_user.username == 'testuser'
assert retrieved_user.email == 'test@example.com'
```

In the preceding test snippet, we created a new user with a specific `username`, `password`, and `email address`. Then, we added the user to the database and commited the changes. Finally, we retrieved the user from the database using a query and asserted that the retrieved user matches the original user in all fields. This test ensures that new users can be successfully created and saved to the database.

Input validation unit test

Let's test that the registration form validates user input correctly and returns appropriate error messages for invalid input:

```
def test_user_registration_input_validation(client, db):
    # Attempt to register a new user with an invalid
      username
    response = client.post('/register',
        data={'username': 'a'*51,
            'password': 'testpassword',
                'email': 'test@example.com'})
    # Assert that the response status code is 200 OK
    assert response.status_code == 200
    # Assert that an error message is displayed for the
      invalid username
    assert b'Invalid username. Must be between 1 and 50
        characters.' in response.data
    # Attempt to register a new user with an invalid email
      address
    response = client.post('/register',
        data={'username': 'testuser',
            'password': 'testpassword',
                'email': 'invalid-email'})
    # Assert that the response status code is 200 OK
    assert response.status_code == 200
    # Assert that an error message is displayed for the
      invalid email address
    assert b'Invalid email address.' in response.data
```

```
        # Attempt to register a new user with a password that
          is too short
        response = client.post('/register',
            data={'username': 'testuser',
                'password': 'short',
                    'email': 'test@example.com'})

        # Assert that the response status code is 200 OK

        assert response.status_code == 200
        # Assert that an error message is displayed for the
          short password
        assert b'Password must be at least 8 characters long.'
            in response.data
```

In the preceding test, we simulated attempts to register a new user with various invalid inputs, such as an invalid username, email address, or password properties that are too short. We sent POST requests to the '/register' endpoint with this invalid input data and asserted that the response status code was 200 OK, indicating that the registration form was submitted successfully, but with errors.

Then, we asserted that the appropriate error messages were displayed on the page for each invalid input. This test ensures that the registration form correctly validates the user input and returns appropriate error messages for invalid input.

Next, we will examine unit testing the login component.

Unit-testing user login

Unit testing user login involves testing the functionality of the code responsible for authenticating a user who attempts to log into an application. This typically involves verifying that user credentials are correct and that the appropriate response is returned based on whether the authentication was successful or not.

Unit testing in this context can help ensure that the login process is reliable and secure, with appropriate error handling for invalid login attempts. Additionally, unit testing can help identify potential vulnerabilities in the login process, such as injection attacks or password-guessing attempts.

User with valid credentials unit test

Let's test that a user with valid credentials can successfully log in and access the application:

```
def test_user_login(client, user):
    # Login with valid credentials
    response = client.post('/login',
```

```
        data={'username': user.username,
            'password': user.password},
        follow_redirects=True)
    # Check that the response status code is 200 OK
    assert response.status_code == 200
    # Check that the user is redirected to the home page
        after successful login
    assert b'Welcome to the application!' in response.data
```

In the preceding test, we're using the client fixture to simulate a user logging in by sending a POST request to the login endpoint with valid credentials. We're also using the user fixture to create a test user with valid credentials. After sending the login request, we check that the response status code is 200 OK and that the user is redirected to the home page, which indicates that the login was successful.

User with invalid credentials unit test

Let's test that a user with invalid credentials cannot log in and receives an appropriate error message:

```
def test_login_invalid_credentials(client):
    # Try to log in with invalid credentials
    response = client.post('/login',
        data={'username': 'nonexistentuser',
        'password': 'wrongpassword'})
    # Check that the response status code is 401
        Unauthorized
    assert response.status_code == 401
    # Check that the response contains the expected error
        message
    assert b'Invalid username or password' in response.data
```

In the preceding test, we are trying to log in with a username and password that are not valid, and we expect the server to respond with a 401 Unauthorized status code and an error message indicating that the credentials were invalid.

Testing SQL injection attacks

Let's test that the code is properly validating user input to prevent SQL injection attacks:

```
def test_sql_injection_attack_login(client):
    # Attempt to login with a username that contains SQL
        injection attack code
    response = client.post('/login',
        data={'username': "'; DROP TABLE users; --",
            'password': 'password'})
```

```
    # Check that the response status code is 401
      Unauthorized
    assert response.status_code == 401
    # Check that the user was not actually logged in
    assert current_user.is_authenticated == False
```

In the preceding test, we are attempting to use SQL injection attack code as the username input in the login form. The test checks that the response status code is 401 Unauthorized, indicating that the attack was not successful, and the user was not logged in.

It also checks that the current_user.is_authenticated attribute is False, confirming that the user is not authenticated. This test helps ensure that the code is properly validating user input to prevent SQL injection attacks.

Testing for password strength

Let's test that the code is properly validating user passwords to ensure they meet the minimum complexity requirements (for example, a minimum length, the requirement of special characters, and so on):

```
def test_password_strength():
    # Test that a password with valid length and characters
      is accepted
    assert check_password_strength("abc123XYZ!") == True

    # Test that a password with an invalid length is rejected
    assert check_password_strength("abc") == False

    # Test that a password without any special characters
      is rejected
    assert check_password_strength("abc123XYZ") == False

    # Test that a password without any lowercase letters is
      rejected
    assert check_password_strength("ABC123!") == False

    # Test that a password without any uppercase letters is
      rejected
    assert check_password_strength("abc123!") == False

    # Test that a password without any numbers is rejected
    assert check_password_strength("abcXYZ!") == False
```

In the preceding test, `check_password_strength()` is a function that takes a password string as input and returns `True` if it meets the minimum complexity requirements and `False` otherwise. This unit test verifies that the function works as expected by testing various scenarios.

With the use of a testing framework, Pytest, and writing effective unit tests, developers can catch bugs and defects early on, reducing the risk of errors in production and improving the overall quality and reliability of their code base.

> **Note**
>
> The preceding tests assumed that you have a Flask application set up with routes for user registration and login, as well as a SQLAlchemy database with a user model. We also assume that you have a test client configured with Pytest's Flask test client fixture (client).

Next, we will look at testing JSON APIs to make sure that the API endpoints work as expected.

Testing JSON APIs

Testing JSON APIs is an essential part of developing any web application that communicates with external clients. APIs provide a simple and flexible way to exchange data between the server and the client. APIs are critical to ensure that the APIs work as expected before they are exposed to external users.

Unit-testing JSON APIs involves verifying that the API endpoints return the expected results for different types of input data and handling error cases. Additionally, it's essential to ensure that the API follows industry-standard protocols and is secure against common web vulnerabilities. In this way, developers can ensure the reliability and security of the web application and minimize the risk of errors or security breaches.

Let's go through a test suite with four tests – `test_get_all_speakers`, `test_create_speaker`, `test_update_speaker`, and `test_delete_speaker`:

```python
import pytest
import requests

# Define the base URL for the speakers API
BASE_URL = 'https://localhost:5000/v1/api/speakers/'

def test_get_all_speakers():
    # Send a GET request to the speakers API to retrieve
      all speakers
    response = requests.get(BASE_URL)

    # Check that the response has a status code of 200 OK
    assert response.status_code == 200
```

```
# Check that the response contains a JSON object with a
  list of speakers
assert isinstance(response.json(), list)
```

The preceding test, `test_get_all_speakers`, sends a GET request to the speakers API to retrieve all speakers and then checks that the response has a status code of 200 OK and contains a JSON object with a list of speakers.

Testing speaker data creation

The following test, `test_create_speaker`, defines a speaker data object to be created, sends a POST request to the Speakers API to create a new speaker using this data, and then checks that the response has a status code of 201 CREATED and contains a JSON object with the newly created speaker data:

```
def test_create_speaker():
    # Define the speaker data to be created
    speaker_data = {
        'name': 'John Darwin',
        'topic': 'Python',
        'email': 'john@example.com',
        'phone': '555-555-5555'
    }

    # Send a POST request to the speakers API to create a
      new speaker
    response = requests.post(BASE_URL, json=speaker_data)

    # Check that the response has a status code of 201
      CREATED
    assert response.status_code == 201

    # Check that the response contains a JSON object with
      the newly created speaker data
    assert response.json()['name'] == 'John Darwin'
    assert response.json()['topic'] == 'Python'
    assert response.json()['email'] == 'john@example.com'
    assert response.json()['phone'] == '555-555-5555'
```

Updating the speaker data object

The following test code, `test_update_speaker`, defines a speaker data object to be updated, sends a PUT request to the Speakers API to update the speaker with `id` 1 using this data, and then checks that the response has a status code of 200 for a successful update:

```python
def test_update_speaker():
    # Define the speaker data to be updated
    speaker_data = {
        'name': 'John Doe',
        'topic': 'Python for Data Science',
        'email': 'johndoe@example.com',
        'phone': '555-555-5555'
    }

    # Send a PUT request to the speakers API to update the
      speaker data
    response = requests.put(BASE_URL + '1',
        json=speaker_data)

    # Check that the response has a status code of 200 OK
    assert response.status_code == 200

    # Check that the response contains a JSON object with
      the updated speaker data
    assert response.json()['name'] == 'John Darwin'
    assert response.json()['topic'] == 'Python for Data
        Science'
    assert response.json()['email'] == 'john@example.com'
    assert response.json()['phone'] == '555-555-5555'
```

Testing the deletion of the speaker data object

The following code snippet sends a DELETE request to the Speakers API to delete the speaker with ID 1. The test function checks that the response has a status code of 204 NO CONTENT. If the speaker with ID 1 is successfully deleted from the API, the response from the API should have a status code of 204 NO CONTENT. If the speaker is not found or if there is an error in the delete request, the response status code will be different, and the test will fail:

```python
def test_delete_speaker():
    # Send a DELETE request to the speakers API to delete
        the speaker with ID 1
```

```
response = requests.delete(BASE_URL + '1')

# Check that the response has a status code of 204 NO
  CONTENT
assert response.status_code == 204
```

At this point, you might be wondering, why do we need to invest time and resources into rectifying bugs once they've emerged in our application when it's entirely possible to proactively forestall their occurrence from the outset?

Next, we will discuss TDD using Flask as a significant proactive approach to software development!

Test-driven development with Flask

TDD is a software development approach where you write automated tests before writing the actual code. The process involves writing a test case for a specific feature or functionality and then writing the minimum amount of code necessary to make the test pass. Once the test passes, you write additional tests to cover different edge cases and functionality until you have fully implemented the desired feature.

Using Flask with an attendee endpoint as a case study, the TDD process might look like this:

1. **Define the feature**: The first step is to define the feature you want to implement. In this case, the feature is an endpoint that allows users to view a list of attendees for an event.

2. **Write a test case**: Next, you must write a test case that defines the expected behavior of the endpoint. For example, you might write a test that checks that the endpoint returns a JSON response with a list of attendees.

3. **Run the test**: You then run the test, which will fail since you haven't implemented the endpoint yet.

4. **Write the minimum amount of code**: You write the minimum amount of code necessary to make the test pass. In this case, you would write the code for the attendee endpoint.

5. **Run the test again**: Then, you must run the test again, which should now pass since you've implemented the endpoint.

6. **Write additional tests**: Finally, you must write additional tests to cover different edge cases and functionality. For example, you might write a test that checks that the endpoint returns a 404 error if the event doesn't exist.

Now, let's implement the attendee's endpoint using the TDD approach, starting with a failed test case since we haven't implemented the endpoint yet.

Defining the feature

The first step is to define the feature you want to implement. In this case, the feature is an endpoint that allows users to view a list of attendees for an event.

Writing a failed test case

The next step is to write a test case that checks that the attendee endpoint returns the expected data. This test should fail initially since we haven't implemented the endpoint yet.

Create test_attendees.py inside the tests directory and add the following code to bizza/backend/tests/test_attendees.py:

```python
from flask import Flask, jsonify
import pytest
app = Flask(__name__)


@pytest.fixture
def client():
    with app.test_client() as client:
        yield client


def test_attendees_endpoint_returns_correct_data(client):
    response = client.get('/events/123/attendees')
    expected_data = [{'name': 'John Darwin',
        'email': 'john@example.com'},
            {'name': 'Jane Smith',
                'email': 'jane@example.com'}]
    assert response.json == expected_data
```

Implementing the minimal amount of code to pass the test

Now, we can implement the attendee endpoint function to return the hardcoded data. This is the minimal amount of code necessary to make the test pass:

```python
# Define the attendee endpoint
@app.route('/events/<int:event_id>/attendees')
def get_attendees(event_id):
    # Return a hardcoded list of attendees as a JSON
      response
    attendees = [{'name': 'John Darwin',
        'email': 'john@example.com'},
            {'name': 'Jane Smith',
                'email': 'jane@example.com'}]
    return jsonify(attendees)
```

Running the test and ensuring it passes

Run the test again to ensure that it now passes:

```
$ pytest test_attendees.py


- - - - - - - - - - - - - - - - - - - - - - - - - - - - - - - - - - - - - - - - -
Ran 1 test in 0.001s

OK
```

Refactoring the code

Now that we have a passing test, we can refactor the code to make it more maintainable, efficient, and readable. For example, we could replace the hardcoded data with data retrieved from a database or external API.

Writing additional tests

Finally, we can write additional test cases to ensure that the endpoint behaves correctly in different scenarios. For example, we might write tests to ensure that the endpoint handles invalid input correctly, or that it returns an empty list if no attendees are found for a given event.

With the TDD process, you can ensure that your code is thoroughly tested and that you've implemented all the desired functionalities. This approach can help you catch bugs early in the development process and make it easier to maintain and refactor your code in the future.

So far, we have discussed TDD as a software development approach where tests are created before the actual code implementation. This approach encourages developers to write tests that define the expected behavior of their code and then write the code itself to make the tests pass. Next, we will delve into the realm of exception handling in a test suite in Flask.

Handling exceptions

Handling exceptions with unit testing is a software development technique that involves testing how a piece of code handles different types of exceptions that may occur during runtime. Exceptions can be triggered by a variety of factors, such as invalid input, unexpected input, or issues with the environment in which the code is running.

Unit testing is the practice of writing small, automated tests to ensure that individual units of code are working as expected. When it comes to handling exceptions, unit tests can help ensure that the code responds appropriately to various error conditions. As a developer, you need to test that your code can handle exceptions gracefully. You can simulate these error conditions in a controlled environment so that you have more confidence in your code's ability to handle exceptions that may occur.

For instance, in the case of a Flask application with an `attendees` endpoint, you may want to test how the application handles requests for events with no attendees. By writing a unit test that sends a request to the endpoint with an event that has no attendees, we can ensure that the application returns the appropriate error response code and message, rather than crashing or providing an inaccurate response.

Let's dive into a code implementation of how you can handle exceptions for attendees' endpoints:

```python
from flask import Flask, jsonify
app = Flask(__name__)
class Event:
    def __init__(self, name):
        self.name = name
        self.attendees = []

    def add_attendee(self, name):
        self.attendees.append(name)

    def get_attendees(self):
        if not self.attendees:
            raise Exception("No attendees found for event")
        return self.attendees

@app.route('/event/<event_name>/attendees')
def get_attendees(event_name):
    try:
        event = Event(event_name)
        attendees = event.get_attendees()
    except Exception as e:
        return jsonify({'error': str(e)}), 404
    return jsonify(attendees)
```

In the preceding implementation, we've added a custom exception to the `Event` class called `Exception("No attendees found for event")`. In the `get_attendees` method, if there are no attendees, we raise this exception. In the Flask endpoint function, we wrap the `Event` instantiation and the `get_attendees` call in a `try/except` block.

If an exception is raised, we return a JSON response with the error message and a `404` status code to indicate that the requested resource was not found.

Let's examine the test function:

```python
def test_get_attendees_empty():
    event_name = 'test_event'
    app = create_app()
```

```
    with app.test_client() as client:
        response =
            client.get(f'/event/{event_name}/attendees')
        assert response.status_code == 404
        assert response.json == {'error': 'No attendees
            found for event'}


def test_get_attendees():
    event_name = 'test_event'
    attendee_name = 'John Doe'
    event = Event(event_name)
    event.add_attendee(attendee_name)
    app = create_app()
    with app.test_client() as client:
        response =
            client.get(f'/event/{event_name}/attendees')
        assert response.status_code == 200
        assert response.json == [attendee_name]
```

In the first test function, `test_get_attendees_empty()`, we expect the endpoint to return a 404 status code and an error message JSON response because there are no attendees for the event. In the second test, `test_get_attendees()`, we add an attendee to the event and expect the endpoint to return a 200 status code and a JSON response containing the attendee's name.

When you test for expected exceptions and handle them gracefully in your code, you can ensure that your application behaves as expected and provides helpful error messages to users when needed.

Summary

Unit testing, as a crucial aspect of Flask application development, ensures the reliability and functionality of application software. In this chapter, we learned how to structure and implement effective unit tests for various components of a Flask application. We explored how Pytest simplifies testing processes and enhances the productivity of developers.

This chapter covered the fundamentals of Pytest, including its introduction, setup process, basic syntax, and features. We discovered the importance of the setup and teardown methods, which help create a controlled testing environment and ensure the proper disposal of resources after each test case.

By applying these techniques, we were able to create more robust and isolated unit tests that mirror real-world scenarios. Furthermore, we provided guidelines on how to write unit tests, test JSON APIs, apply TDD, and handle exceptions in Flask applications. With the adoption of these practices, developers can improve the overall quality of their Flask applications and minimize the risk of errors and bugs.

As we move forward and wrap up our journey of building robust and scalable Flask applications, the next chapter will dive into the world of containerization and deployment. We will explore how to containerize Flask applications, allowing us to replicate development environments and effortlessly deploy our applications to various platforms.

We will also delve into deploying Flask applications to cloud services, harnessing the power of platforms such as Docker and AWS for efficient and scalable deployment.

16

Containerization and Flask Application Deployment

After a long journey, we've reached the last chapter! We're thrilled beyond words! Right now, we are about to embark on the last lap of showcasing our full stack web application to the world. In today's modern software development sphere, the pace of containerization adoption is rapidly increasing.

According to Gartner's predictions, the adoption of containerized applications in production will increase significantly, with more than 75% of global organizations expected to utilize them by 2022, a notable increase from the less than 30% reported in 2020 (`https://www.gartner.com/en/documents/3985796`).

Containerization and the deployment of software applications have become essential skills needed for developers to stay modern and in demand. Developers who have the skills and knowledge to containerize and deploy software applications are better equipped to meet the demands of modern software development practices, stay up to date with industry trends, and remain competitive in the job market.

Containerization allows developers to package applications and required dependencies into a standardized and portable container that can run consistently across different computing environments. And, of course, deployment ensures that your application gets to the production environment, where it can be used by end users.

In this chapter, we will discuss containerization as a revolution changing the information technology industry. We will touch on the significance and benefits of containerization in software development, as well as exploring the issues it tackles.

Furthermore, we will delve into one of the containerization platforms in the software development industry, called **Docker**. We will introduce Docker and use it to containerize both the React frontend and the Flask backend. We will discuss Docker's benefits and why it is popular among developers.

By the end of the chapter, you will understand the importance of containerization in modern software development, and you will be able to package React and Flask applications into containers ready to be shipped and shared.

Finally, you will learn how to use **Amazon Web Services** (**AWS**) Elastic Beanstalk to deploy React-Flask applications leveraging the AWS fully managed cloud platform, which allows developers to deploy, manage, and scale their web applications and services with ease.

In this chapter, we'll cover the following topics:

- What is containerization?
- Introducing Docker
- Dockerizing React and Flask applications
- Understanding AWS Elastic Container Registry
- Using `docker-compose`
- Deploying React and Flask applications to AWS Elastic Beanstalk

Technical requirements

The complete code for this chapter is available on GitHub at: `https://github.com/PacktPublishing/Full-Stack-Flask-and-React/tree/main/Chapter16`.

Due to page count constraints, some of the long code blocks have been shortened. Please refer to GitHub for the complete code.

What is containerization?

Containerization is a software development practice that involves packaging an application and required dependencies into a self-contained unit called a **container**. A container is an isolated and lightweight runtime environment that provides a consistent and reproducible way to run an application across different computing environments.

Let's say you have developed a web application using the Flask framework on your local machine running on MacOS. You want to deploy this application to a server running Ubuntu Linux in a production environment. However, there may be differences in the versions of the operating system, dependencies, or other system configurations that could affect the behavior of your application.

By packaging your Flask application and all the required dependencies into a container, you can ensure that the application runs consistently and reliably across different computing environments. The container will provide an isolated and lightweight runtime environment that encapsulates the application and related dependencies, ensuring that it runs consistently regardless of the underlying system configurations.

Containers are like **virtual machines (VMs)** in that they provide a way to isolate applications from the underlying host operating system. However, while VMs require a complete copy of the host operating system to run, containers only require the minimal runtime components needed to run the application.

Containers utilize a technique called containerization, which employs virtualization at the operating system level. Containerization allows multiple containers to run on the same host operating system, each with its own isolated filesystem, networking, and process space. With containerization, developers can lower the deployment time and cost.

Let's check out a few other benefits of containerization:

- Containers provide a standardized way of packaging applications and required dependencies, which reduces the time and effort required for configuration and setup.

- Containers are portable across different computing environments, allowing for deployment on any system with the same container runtime. This portability approach eliminates the need to create and maintain separate deployment configurations for different environments.

- Containers share the same operating system kernel, enabling more efficient use of system resources compared to traditional virtualization.

- Containers provide isolation between applications and accompanying dependencies, making the conflicts and errors that can arise when running applications on a shared infrastructure obsolete.

In a nutshell, containers are lightweight, self-containing to run applications, portable, and efficient, and can be easily replicated and scaled as needed.

Although there are several containerization technologies to choose from, we will specifically discuss Docker in the next section. Before exploring Docker in depth, let's take a brief look at some of the other containerization tools and platforms available in the software industry:

- **Kubernetes**: An open source container orchestration system that automates deploying, scaling, and managing containerized applications

- **Apache Mesos**: An open source platform for managing and deploying containerized applications and big data services

- **LXC/LXD**: A containerization solution that uses lightweight VMs to provide isolation and resource management

- **CoreOS rkt**: A container runtime that provides security, simplicity, and speed to the container environment

- **OpenVZ**: An open source containerization platform that provides container-based virtualization for Linux

- **AWS Elastic Container Service (ECS)**: A fully managed container orchestration service provided by AWS

- **Google Kubernetes Engine (GKE)**: A fully managed Kubernetes service provided by Google Cloud Platform

As the demand for scalable and efficient software deployment grows, more and more developers are going to turn to Docker as a solution. In the next section, we'll explore the basics of Docker and how it can help you streamline your development workflow.

Introducing Docker

Docker is a popular platform for developing, packaging, and deploying applications in containers. Before Docker's invention, software developers had to deal with the problem of software dependencies, which meant that the software would work well on one computer but fail to work on another system.

Software developers would create programs on their computers, but when they tried to share them with other people, things often went wrong. Programs that worked perfectly on one computer might not have worked on another because of differences in the operating system, software versions, configuration files, or other system-related factors.

To solve this problem, a group of developers in 2013 released a tool called Docker. Docker lets developers package programs and all the necessary dependencies into something called a **Docker image**. A Docker image is a read-only template that contains the instructions for creating a Docker container. A Docker image includes the application code, runtime, libraries, dependencies, and configurations needed to run the application.

With Docker, developers can create a Docker image for their programs and share it with others. A Docker container is a runnable instance of a Docker image. A Docker container is a lightweight, isolated, and portable environment that can run on any system that supports Docker. This means that the program will run the same way on every computer, which makes it much easier to share and deploy.

Developers can create a Docker image by writing a **Dockerfile**, which is a text file that contains the instructions for building a Docker image. The Dockerfile specifies the base image, adds the necessary packages and files, and sets the configuration options for the application.

Once you have built your application Docker image, you might want to ship to production or send it to other developers. To achieve this, you can use a Docker registry, which is a central repository for storing and distributing Docker images. Docker Hub is the most popular public registry, but you can also set up your own private registry for your organization. In the course of this chapter, we will store the book project Docker images in AWS **Elastic Container Registry** (**ECR**).

Docker Compose is another tool of interest in the Docker ecosystem. Docker Compose is a tool for defining and running multi-container Docker applications. Docker Compose uses a YAML file to define the services, networks, and volumes needed to run the application. In the subsequent section, we will discuss Docker Compose in great detail. Next, we will explore how we can containerize a simple Flask application.

Creating a Flask application

Now, let's demonstrate with a simple Flask application the process of containerization using Docker:

1. Download Docker from `https://docs.docker.com/get-docker/` and install Docker on your system.

Figure 16.1 – The download page of Docker

2. Select the appropriate computer OS for your Docker platform.

3. Once the Docker installation is complete, test it in your Terminal with the following command:

    ```
    Docker --version
    ```

    Alternatively, you can use `Docker version` and you will get the following output:

    ```
    Command Prompt

    C:\>Docker --version
    Docker version 20.10.12, build e91ed57

    C:\>
    ```

Figure 16.2 – The command to verify the Docker installation

4. Now that Docker is installed on your computer, run the `mkdir bizza-docker` command to create a new working directory for deploying a Flask application using a Docker container. Then, enter `cd bizza-docker`:

Figure 16.3 – The creation of a Docker working directory

Let's create a virtual environment for the new Flask Docker application.

5. Run `python -m venv venv` in the Terminal to install a virtual environment.

6. Activate the virtual environment with these commands:

    *   **Windows**: `venv\Scripts\activate`

    *   **MacOS/Linux**: `source venv/bin/activate`

7. Inside the Docker project directory, `bizza-docker/`, create an `app.py` file and add the following code snippet to run a simple Flask application:

```
from flask import Flask
app = Flask(__name__)

@app.route('/')
def index():
    return "Bizza Web App Dockerization!"

if __name__ == "__main__":
    app.run(host='0.0.0.0', port=5001, debug=True)
```

The preceding code runs a simple Flask app showing **Bizza Web App Dockerization!** in the browser.

8. Create a `.flaskenv` file inside `bizza-docker/` and add the following code snippet:

```
FLASK_APP = app.py
FLASK_DEBUG = True
```

9. Now, run the Flask app with `flask run` in the Terminal and you will get the following output:

Figure 16.4 – Testing the Flask application

Now that the Flask app is working, let's create a Flask application `requirements.txt` file to be able to reproduce the dependencies for this simple application.

10. Run the `pip freeze > requirements.txt` command and you will get the following output:

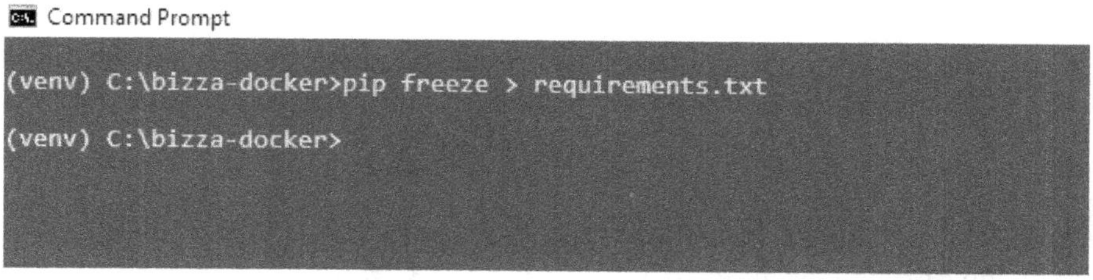

```
Command Prompt

(venv) C:\bizza-docker>pip freeze > requirements.txt

(venv) C:\bizza-docker>
```

Figure 16.5 – The requirements.txt file for the Flask dependencies

The following block displays the content of the `requirements.txt` file:

```
blinker==1.6.2
click==8.1.3
colorama==0.4.6
Flask==2.3.2
itsdangerous==2.1.2
Jinja2==3.1.2
MarkupSafe==2.1.2
python-dotenv==1.0.0
Werkzeug==2.3.3
```

We now have all the resources to build the Docker image.

Creating a Dockerfile

A Dockerfile defines the container image for a Flask application. We are going to create a Dockerfile that uses the official Python 3.8 image as the base image, installs Flask and its dependencies, and copies the Flask application code into the container.

In the `bizza-docker` directory, create a Dockerfile file – make sure the capital *D* is used in creating the Dockerfile file. Don't worry about why; this is a convention:

```
FROM python:3.8.2-alpine
WORKDIR /packt-bizza-docker
ADD . /packt-bizza-docker
COPY requirements.txt .
RUN pip install --no-cache-dir -r requirements.txt
RUN pip3 install -r requirements.txt
COPY . .
```

```
ENV FLASK_APP=app.py
ENV FLASK_ENV=development
EXPOSE 5001
CMD ["python3", "app.py"]
```

To simplify the deployment process and ensure consistent environments across different stages of development, testing, and production, let's examine the anatomy of the preceding Dockerfile.

The preceding code is a Dockerfile used to build a Docker image for a Python Flask web application:

- `FROM python:3.8.2-alpine`: This specifies the base image to use for building the Docker image. In this case, the base image is `python:3.8.2-alpine`, which is a lightweight version of Python 3.8.2 optimized for running in Alpine Linux.

- `WORKDIR /packt-bizza-docker`: This sets the working directory of the Docker container to `/packt-bizza-docker`. All subsequent commands in the Dockerfile will be executed relative to this directory.

- `ADD . /packt-bizza-docker`: This line copies all the files and directories in the current directory into the `/packt-bizza-docker` directory in the Docker container.

- `COPY requirements.txt .`: This copies the `requirements.txt` file from the current directory to the root directory of the Docker container.

- `RUN pip install --no-cache-dir -r requirements.txt`: This installs the Python dependencies specified in the `requirements.txt` file using `pip`. The `--no-cache-dir` option ensures that `pip` does not cache the downloaded packages.

- `RUN pip install -r requirements.txt`: This line installs the Python dependencies specified in the `requirements.txt` file using `pip3`.

- `COPY . .`: This copies all the files and directories in the current directory to the root directory of the Docker container. This includes the Flask application code.

- `ENV FLASK_APP=app.py`, `ENV FLASK_ENV=development`: This sets the environment variables for the Flask application. `FLASK_APP` specifies the name of the main Flask application file (in this case, `app.py`). `FLASK_ENV` sets the Flask environment to development mode.

- `EXPOSE 5001`: This exposes port `5001` of the Docker container.

- `CMD ["python3", "app.py"]`: This specifies the command to run when the Docker container is started. In this case, it runs the `app.py` file using Python 3.

Next, we will build the Docker image from the preceding defined `Dockerfile`.

Building the Docker image

With the Dockerfile defined, you can build a Docker image of the Flask application. This image contains all the dependencies and configuration files required to run the application.

To construct the Docker image, execute the following command in the Terminal from within the `bizza-docker` directory that contains the Dockerfile:

```
docker build -t packt-bizza-docker .
```

We will get the following output:

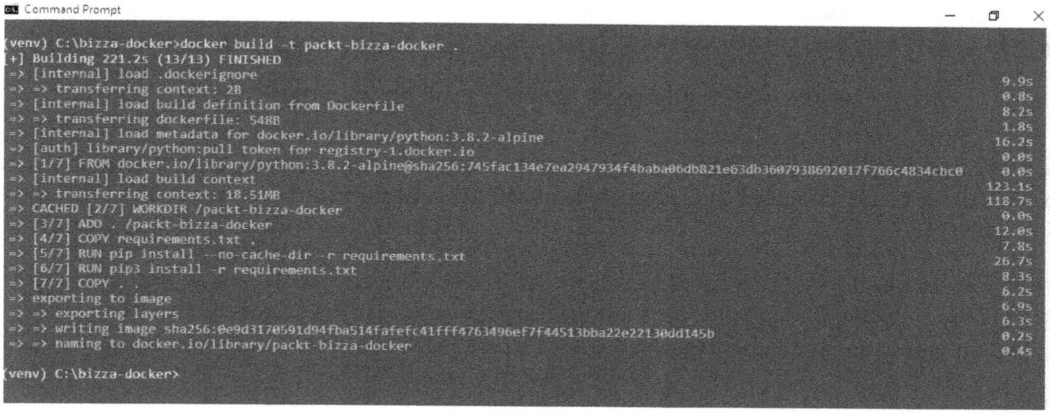

Figure 16.6 – Output of the docker build command

The preceding command will build the Docker image using the Dockerfile present in the current directory. The resulting image will be tagged as `packt-bizza-docker`. Now, let's proceed to the next step and launch the container to make the simple Flask application functional.

Running the Docker container

After building the Docker image, you can run a Docker container from the image. This container provides a lightweight, isolated, and portable environment for running the Flask application.

To run the Docker container, use the following command:

```
docker run -d -p 5001:5001 packt-bizza-docker .
```

We will get the following output:

Figure 16.7 – Output of docker run in detached mode

The preceding command will run the container in detached mode (-d) and perform port mapping (-p) by mapping the host port 5001 to the container port 5001. The container will be based on the packt-bizza-docker image. Alternatively, you can run the command without the -d flag to launch the container in a non-detached mode, as shown in the following figure:

```
(venv) C:\bizza-docker>docker run -p 5001:5001 packt-bizza-docker
* Serving Flask app 'app'
* Debug mode: on
WARNING: This is a development server. Do not use it in a production deployment. Use a production WSGI server instead.
* Running on all addresses (0.0.0.0)
* Running on http://127.0.0.1:5001
* Running on http://172.17.0.2:5001
Press CTRL+C to quit
* Restarting with stat
* Debugger is active!
* Debugger PIN: 815-754-813
```

Figure 16.8 – Output of docker run in a non-detached mode

The preceding docker run command allows us to access the Flask application running inside the Docker container. You need to expose the ports on the container to the host machine with -p 5001:5001 ..

Now that we have the Docker container running, we can test the Flask application by accessing it through a web browser or using a command-line tool such as curl - http://127.0.0.1:5001. Make sure that the application is functioning as expected and that all the dependencies are working correctly.

Finally, you can push the Docker image to a Docker registry such as Docker Hub or AWS ECS. This makes it easy to share the image with other developers or deploy it to production environments.

To stop a running Docker container, you can use the docker stop command followed by the *container ID* or *name.*

For example, if the container ID is c2d8f8a4b5e3, you can stop the container using the docker stop c2d8f8a4b5e3 command, as shown in the following figure:

```
(venv) C:\bizza-docker>docker stop 738457153daeb51d0e21b7223e386632547fdb6b43f17afdcbf5c0c8ab9ba453
738457153daeb51d0e21b7223e386632547fdb6b43f17afdcbf5c0c8ab9ba453

(venv) C:\bizza-docker>_
```

Figure 16.9 – Output of the docker stop command

And if you don't know the container ID or name, you can use the docker ps command to list all running containers and their details, including the ID and name. Once you have identified the container that you want to stop, you can use the docker stop command as described earlier.

Let's glance at another important Docker command: docker container prune.

The `docker container prune` command is used to remove stopped containers and free up disk space. When you run a Docker container, the container consumes system resources such as memory and CPU cycles. When you stop a container, those resources are freed up, but the container still exists on your system. With time, if you run multiple containers, you may be housing many stopped containers, which can take up significant disk space on your system.

Running the `docker container prune` command is a simple way to remove all stopped containers and reclaim disk space. This `docker container prune` command will prompt you to confirm that you want to remove the containers before proceeding, so make sure you review the list of containers carefully before confirming.

It's important to note that the `docker container prune` command will only remove stopped containers. If you have any running containers, they will not be affected by this command.

Next, we will discuss the process of dockerizing React and Flask applications. We will use the full stack Bizza web application as a case study.

Dockerizing React and Flask applications

Dockerizing web applications allows developers to set up a consistent development environment across different machines. Dockerizing tools reduce the time and effort required to set up a new development environment. With Docker, developers can easily replicate the production environment on their local machines, test their code, and debug any issues before deploying it.

In this section, we will dockerize working applications for React and Flask, and make them ready to be shipped for production.

Let's start with the React.

Bizza frontend application with React

Once you have created your React application, the initial step toward making it accessible to internet users is to build the application. Building a React application is an essential step in the development process to ensure that the application is optimized for production and performs as expected.

The building process takes the source code of a React project and transforms it into a production-ready format that can be deployed and served to users:

1. Let's download the *Bizza* app directory from the GitHub repo – `https://github.com/PacktPublishing/Full-Stack-Flask-and-React/tree/main/Chapter16/bizza/frontend`.
2. To install the dependencies required for the application, navigate to the `bizza/frontend` directory and execute the `npm install` command in the Terminal.

3. To run the frontend application, execute the `npm start` command in the Terminal.

4. Now, let's build the application with the `npm run build` command.

Now that the *bizza* React application has been built, the resulting files can be deployed to a web server or cloud platform and served to users. The eventual build directory is located inside `bizza/frontend/src/build`.

During the build process, the following steps were taken:

1. **Transpiling JavaScript and JSX code**: React applications are typically written in JavaScript and JSX, a syntax extension for JavaScript. However, modern web browsers can only execute JavaScript code. Therefore, before deploying a React application, the JavaScript and JSX code needs to be transpiled into plain JavaScript using a tool such as Babel.

2. **Bundling the code and assets**: React applications often consist of multiple components, modules, and assets such as images, CSS files, and fonts. Bundling involves grouping all the required code and assets into a single file or set of files that can be served to the user.

3. **Optimizing the code and assets**: To improve performance, the bundled code and assets can be optimized by minifying, compressing, or removing unnecessary code.

4. **Creating a build directory**: Once the code and assets are bundled and optimized, the code needs to be placed in a directory that can be served by a web server. This directory is typically named `build` or `dist`.

Now, typically at this stage, the `build` directory contents are deployed to a web server or cloud for end users. However, for the deployment process outlined in this book, you will utilize a Docker functionality known as **multi-stage builds**. A multi-stage build is a feature in Docker that allows you to create a Docker image that consists of multiple stages, where each stage is a self-contained Docker image with a specific purpose.

The purpose of a multi-stage build is to optimize the size and efficiency of Docker images. With a multi-stage build, you can reduce the size of your final Docker image by only including the necessary files and dependencies. This results in faster builds, smaller image sizes, and more efficient use of resources.

The multi-stage build process involves creating multiple Docker images, each with a specific purpose. The first stage of the build typically contains the source code, dependencies, libraries, and other necessary files.

The final stage of the build usually contains only the essential files and dependencies required to run the application, resulting in a smaller and more efficient Docker image. The essence of a multi-stage build is to ensure that the intermediate stages are used to build and compile the application but are not included in the final image.

Right now, let's examine a Dockerfile for the React frontend app that uses multi-stage builds:

```
# Build stage
FROM node:14.17.0-alpine3.13 as build-stage

WORKDIR /frontend
COPY package*.json ./
RUN npm install --production
COPY . .
RUN npm run build

# Production stage
FROM nginx:1.21.0-alpine
COPY --from=build-stage /frontend/build /usr/share/nginx/html
EXPOSE 80
CMD ["nginx", "-g", "daemon off;"]

# Clean up unnecessary files
RUN rm -rf /var/cache/apk/* \
           /tmp/* \
           /var/tmp/* \
           /frontend/node_modules \
           /frontend/.npm \
           /frontend/.npmrc \
           /frontend/package*.json \
           /frontend/tsconfig*.json \
           /frontend/yarn.lock
```

Let's break the preceding Dockerfile image-building instruction down:

- **The build stage**: The first part of the Dockerfile creates a build stage using the Node.js 14.17.0-alpine3.13 image as the base. The Dockerfile sets the working directory to /frontend and copies the package*.json files from the local directory to the image. The npm install --production command is then run to install the production dependencies. Next, the Dockerfile copies the entire project directory to the image and runs the npm run build command to build the React app.

- **Production stage**: The second part of the Dockerfile creates a production stage using the smaller nginx:1.21.0-alpine image as the base. The Dockerfile copies the built React app from the build stage, located at /frontend/build, to the nginx HTML directory at /usr/share/nginx/html.

- **EXPOSE**: The EXPOSE command exposes port 80 to allow communication with the container.

- **CMD**: The CMD command sets the default command to run when the container starts up. In this case, the Dockerfile starts the nginx server in the foreground with the nginx -g 'daemon off;' command.

- **Cleaning up unnecessary files**: Finally, to further optimize the image size, the Dockerfile cleans up unnecessary files, such as the node_modules directory and other configuration files, using the RUN command with the rm command to remove them from the image. This cleaning-up process reduces the overall size of the image, making it faster to deploy.

Now that we have the Docker image of the bizza frontend React app. Let's create the Flask backend Docker image.

Bizza backend application with Flask

In the Flask backend, we are going to create two Docker images. Download the full Flask backend application here: https://github.com/PacktPublishing/Full-Stack-Flask-and-React/tree/main/Chapter16/bizza/backend.

We will create a Docker image for the Flask application and another Docker image for PostgreSQL. While it is possible to fuse the two images into a single Docker image, it is a best practice to separate the concerns for scalability and to reduce the image size.

Let's review the Flask application multi-stage build Dockerfile definition.

The Dockerfile for the Flask application will be stored in the project root directory while a subdirectory named postgres will house the Dockerfile for PostgreSQL:

```
# Build Stage
FROM python:3.8.12-slim-buster AS build
WORKDIR /app
COPY requirements.txt .
RUN pip install --no-cache-dir -U pip==21.3.1 && \
    pip install --no-cache-dir --user -r requirements.txt
COPY . .
# Run Stage
FROM python:3.8.12-slim-buster AS run
WORKDIR /app
COPY --from=build /root/.local /root/.local
COPY --from=build /app .
ENV PATH=/root/.local/bin:$PATH
ENV FLASK_APP=app.py
ENV FLASK_ENV=production
EXPOSE 5001
CMD ["python3", "app.py"]
```

Let's break down the preceding Dockerfile.

This Dockerfile defines a multi-stage build for a Flask application. The Dockerfile has two stages: `build` and `run`.

The first stage, `build`, is responsible for building the application and installing the required dependencies. Right now, let's check what each line of the build stage does:

- `FROM python:3.8.12-slim-buster AS build`: This line sets the base image for the build stage to `python:3.8.12-slim-buster`.

- `WORKDIR /app`: This line sets the working directory to `/app`.

- `COPY requirements.txt .`: This line copies the `requirements.txt` file from the host machine to the `/app` directory in the container.

- `RUN pip install --no-cache-dir -U pip==21.3.1 && \   pip install --no-cache-dir --user -r requirements.txt`: These lines update `pip` to version `21.3.1` and install the Python packages specified in the `requirements.txt` file.

  The `--no-cache-dir` option is used to prevent the installation from using any cached data from previous runs, which helps ensure that the installed packages are up to date and match the versions specified in `requirements.txt`. The `--user` option is used to install the packages to the user's home directory, which helps avoid permission issues.

- `COPY . .`: This line copies the entire application directory from the host machine to the `/app` directory in the container.

- `FROM python:3.8.12-slim-buster AS run`: This represents the start of the second stage.

The second stage, `run`, is responsible for running the application in a production environment. The line sets the base image for the `run` stage to `python:3.8.12-slim-buster`:

- `WORKDIR /app`: This line sets the working directory to `/app`.

- `COPY --from=build /root/.local /root/.local` and `COPY --from=build /app .`: These two lines copy the application directory and the installed packages from the build stage to the `run` stage. The first line copies the installed packages from the build stage to the `/root/.local` directory in the run stage. The second line copies the application directory from the build stage to the `/app` directory in the run stage.

- `ENV PATH=/root/.local/bin:$PATH`, `ENV FLASK_APP=app.py`, and `ENV FLASK_ENV=production`: These three lines set the environment variables for the application. The `PATH` environment variable is updated to include the `/root/.local/bin` directory, which contains the installed packages.

This ensures that the installed packages are available in the system PATH. The FLASK_APP environment variable is set to app.py, which specifies the main application file for Flask to run. The FLASK_ENV environment variable is set to production, which enables features such as better error handling and improved performance.

- EXPOSE 5001: This line exposes port 5001, which is the port that the Flask application will listen on.

- CMD ["python3", "app.py"]: This line specifies the default command to run when the container starts. It runs the app.py file using the python3 command.

Having discussed the Dockerfile for the main Flask application, let's examine the Dockerfile for PostgreSQL.

Here's the Dockerfile for Postgres to create a database image:

```
FROM postgres:13-alpine
ENV POSTGRES_DB=<databse_name>
ENV POSTGRES_USER= <databse_user>
ENV POSTGRES_PASSWORD= <databse_password>
RUN apk add --no-cache --update bash
COPY init.sql /docker-entrypoint-initdb.d/
EXPOSE 5432
```

Let's go through the Postgres Dockerfile:

- FROM postgres:13-alpine: This line specifies the base image for our Docker container, which is postgres:13-alpine. This image is based on the Alpine Linux distribution and includes PostgreSQL version 13.

- ENV POSTGRES_DB=<database_name>, ENV POSTGRES_USER=<database_user>, and ENV POSTGRES_PASSWORD=<database_password>: These three lines set the environment variables for the Postgres container. The POSTGRES_DB variable specifies the name of the database to be created. The POSTGRES_USER variable specifies the username to be created for the database, and the POSTGRES_PASSWORD variable specifies the password for that user.

- RUN apk add --no-cache --update bash: This line copies the init.sql file to the /docker-entrypoint-initdb.d/ directory in the container. This directory is used by the Postgres image to run initialization scripts when the container is first started. In this case, the init.sql file is a script that will create the database and any necessary tables.

- EXPOSE 5432: This line exposes port 5432, which is the default port used by PostgreSQL, to allow connections from outside the container. However, this does not actually publish the port, as this needs to be done at runtime using the docker run or docker-compose commands.

This Postgres Dockerfile can be used to build a Docker image for a Postgres database, which can be used in conjunction with React and Flask application Docker containers to build a complete web application stack.

With the Flask application and Postgres images well defined, we will be pushing the created Docker images to AWS ECR for online storage of Docker images.

Understanding AWS ECR

Amazon ECR is a fully managed Docker registry service that makes it easy to store, manage, and deploy Docker images. Amazon ECR is integrated with Amazon ECS to provide a seamless experience for building, deploying, and managing containerized applications at scale. Amazon ECR is designed to scale to meet the needs of even the most demanding containerized applications. Amazon ECR has security features to protect your container images, including encryption at rest and in transit, and **role-based access control (RBAC)**.

To begin using Amazon ECR, the first step is to create an ECR repository. Please refer to the following screenshot of the Amazon ECR interface.

Click on the **Get Started** button to initiate the repository creation process. This will allow you to establish a dedicated location for storing your Docker images.

Figure 16.10 – AWS ECR

Next, we have a screenshot showcasing a public repository named `packt-bizza-web-app` in Amazon ECR:

Figure 16.11 – The public repository

A **repository** is a logical container for storing your Docker images. Once you have created a repository, you can push your Docker images to the repository. You can then pull your images from the repository to deploy them to your ECS clusters.

Amazon ECR is a powerful tool that can help you to simplify the management of your container images. Interestingly, Amazon ECR is very cost-effective in storing and managing container images.

Using ECR is free; you only pay for the storage and bandwidth that you use. Next, we will use Docker Compose to define and run the React, Flask, and Postgres containers.

Using Docker Compose

Docker Compose is a tool for defining and running multi-container Docker applications. Docker Compose provides a tool to define a set of containers and their relationships to each other, and then run them all with a single command.

With Docker Compose, developers can define the exact configuration of the application's containers, including the images, environment variables, and network settings. This ensures that the application runs consistently across different environments and can be easily reproduced.

The following are a few components of Docker Compose we need to understand before we delve into details for configuration definitions:

- **YAML file**: A YAML file is used to define the configuration of your application's containers. The YAML file specifies the images to use, ports to expose, environment variables, and any other settings that are required.

- **Services**: Each container in your application is defined as a service in the YAML file. Services can depend on each other and can be started and stopped together.

- **Networks**: Docker Compose creates a network for your application, allowing the containers to communicate with each other.

- **Volumes**: Volumes are used to persist data between container runs.

- **Commands**: Docker Compose provides a set of commands to start, stop, and manage your application.

Now, let's create a Docker Compose file that manages the relationship between the React frontend, Flask backend, and PostgreSQL database containers:

1. Inside the main project directory, `bizza/`, create `docker-compose.yaml`.

2. Define services for each container. In the `docker-compose.yaml` file, define a separate service for each container:

```yaml
version: '3'
services:
  frontend:
    image: <your-ecr-repository>/bizza-frontend-react-
      app
    ports:
      - "3000:3000"
  backend:
    image: <your-ecr-repository>/bizza-backend-flask-
      app
    ports:
      - "5000:5000"
    depends_on:
      - db
  db:
    image: <your-ecr-repository>/bizza-postgresql
    environment:
      POSTGRES_USER: <your-db-username>
      POSTGRES_PASSWORD: <your-db-password>
      POSTGRES_DB: <your-db-name>
    ports:
      - "5432:5432"
```

In the preceding code, we define three services: `frontend`, `backend`, and `db`. The `frontend` service runs the *Bizza* frontend React app, the `backend` service runs the *Bizza* backend Flask app, and the `db` service runs the PostgreSQL database.

Now, let's configure networking and dependencies. Use the `ports` and `depend_` on options to configure the network connections between the services. For instance, the frontend service is exposed on port `3000`, the backend service is exposed on port `5000`, and the `db` service is exposed on port

5432. The backend service also depends on the db service, so the backend will start after the db service is running.

Once we've defined the services in the `docker-compose.yaml` file, we can start the containers using the `docker-compose up` command. This will start the containers and connect them to the appropriate network.

With Docker Compose managing the application's containers, we can simplify the process of starting and stopping our application, as well as ensure that all the required components are running correctly and communicating with each other.

Interestingly, Docker Compose is a useful tool for managing containers; however, Docker Compose is more suited to small-scale deployments and development environments. Docker Compose serves the purpose of the *bizza* project, being a small-scale application for learning purposes.

However, AWS Elastic Beanstalk, on the other hand, is designed to handle production-grade workloads and provides many features and benefits that can help simplify the management and scaling of web applications. Regardless, we will pivot the *bizza* application's final deployment on AWS Elastic Beanstalk.

In the next section, we will explore AWS Elastic Beanstalk, a service for deploying and managing applications in the cloud.

Deploying React and Flask applications to AWS Elastic Beanstalk

AWS Elastic Beanstalk is a fully managed AWS cloud service that allows developers to easily deploy and manage web applications and services on AWS. AWS Elastic Beanstalk provides a platform that simplifies the process of deploying and managing web applications on AWS by automatically handling the infrastructure provisioning, load balancing, and scaling of the application.

You can deploy Elastic Beanstalk on a wide range of programming languages and web frameworks, including Node.js, Python, Ruby, and Go. Elastic Beanstalk also integrates with other AWS services such as Amazon RDS, Amazon DynamoDB, and Amazon SNS to provide a complete solution for building and scaling web applications.

With Elastic Beanstalk, developers can easily focus on coding. Once you are ready to deploy your application, you can simply upload your application package or link to a repository, and then choose the appropriate platform and environment for your application. Elastic Beanstalk automatically provisions the required resources and sets up the environment and can also automatically scale the application based on demand.

Also, AWS Elastic Beanstalk provides a range of capabilities and tools that help developers streamline their development workflows, such as **continuous integration and continuous delivery (CI/CD)** pipelines, monitoring and logging tools, and integration with popular development tools such as Git and Jenkins.

Now, let's get started with using Elastic Beanstalk to deploy our application. This guide assumes you have created an AWS account. If not, go to `https://aws.amazon.com/free/` and follow the instructions to create an AWS account. The AWS free tier is enough to deploy this book project:

1. Log in to your AWS account and go to the Amazon ECR console at `https://console.aws.amazon.com/ecr/`.

2. To create an Amazon ECR repository, you can use the following steps:

    I.   Go to the Amazon ECR console.

    II. In the navigation pane, select **Repositories**.

    III. Select **Create repository**.

    IV. In the **Repository name** field, enter a name for your repository.

    V.   In the **Repository type** field, select **Public** or **Private**.

    VI. Select **Create**.

3. Alternatively, you can create an Amazon ECR repository with the following AWS CLI command:

    ```
    aws ecr create-repository --repository-name nameofyourrepository
    ```

    However, to successfully run the preceding command you need to have the following sorted:

    *   Have an AWS account and an IAM user with permissions to create ECR repositories. You can find the link to the permissions JSON file on the GitHub at `https://github.com/PacktPublishing/Full-Stack-Flask-and-React/blob/main/Chapter16/bizza/Deployment/ecr-permissions.json`

    *   Have AWS CLI installed and configured with your AWS credentials.

4. Next, we need to push the Docker images to the Amazon ECR repository. To push the bizza application Docker images to the Amazon ECR repository, follow these steps:

    I.   On the command line, navigate to the directory that contains each of the applications' Dockerfile. Build the Docker image with the following command:

    ```
    docker build -t <image-name> .
    ```

    II. Then, tag your image with the following command:

    ```
    docker tag <docker_image_name>:<tag_name> <AWS_ACCOUNT_ID>.dkr.
    ecr.<region>.amazonaws.com/<AWS_REPOSITORY_NAME>:<tag_name>
    ```

    III. Push each of the Docker images to the Amazon ECR repository. Inside your project directory, run `docker login` and enter docker login credentials. Once done, run the `aws configure` command to log in to AWS as well.

IV.   Once you are logged in to both Docker and AWS in your terminal, run the following command:

```
aws ecr get-login-password --region <region> | docker login
--username AWS --password-stdin AWS_ACCOUNT_ID.dkr.ecr.<region>.
amazonaws.com
```

Let's go over the aspects of the preceding command:

*   `aws ecr get-login-password:`- This command retrieves an authentication token from ECR.

*   `--region <region>`: This specifies the region where the ECR registry is located. If you do not know where your ECR repository is located, run the following command: `aws ecr describe-repositories –repository-names nameofyourrepository.`

*   `|`: This is the pipe operator. It tells the shell to take the output of the first command and pass it as input to the second command.

*   `docker login`: This command logs you in to a Docker registry.

*   `--username AWS`: This specifies the username to use when logging in to the registry.

*   `--password-stdin:`- This tells the Docker CLI to read the password from standard input.

*   `<AWS_ACCOUNT_ID>.dkr.ecr.<region>.amazonaws.com:`- This is the registry ID of the ECR registry you want to log in to.

V.   Enter `docker push <account-id>.dkr.ecr.<region>.amazonaws.com/<nameof yourrepository:<tag_name>>` in each of the project component directories.

5.   To create an Elastic Beanstalk environment, you can use the following steps:

I.   Go to the Elastic Beanstalk console at `https://console.aws.amazon.com/elasticbeanstalk`.

II.   In the navigation pane, select **Create environment**.

III.   In the **Platform** section, select **Docker**.

IV.   In the **Application** code section, select **Use an existing application**.

V.   In the **Application code repository** field, enter the URL of your Docker image repository.

VI.   In the **Application name** field, enter a name for your environment.

VII.   Select **Create environment**.

6. To configure the Elastic Beanstalk environment to use the Amazon ECR repository, you can use the following steps:

    I.   In the Elastic Beanstalk console, select the name of your environment.

    II.   In the navigation pane, select **Configuration**.

    III. In the **Software** section, select **Docker**.

    IV.   In the **Repository URL** field, enter the URL of your Amazon ECR repository.

    V.   Select **Save**.

7. To deploy the application to the Elastic Beanstalk environment, you can use the following steps:

    I.   In the Elastic Beanstalk console, select the name of your environment.

    II.   In the navigation pane, select **Deploy**.

    III. In the **Deployment method** section, select **One-click deploy**.

    IV.   Select **Deploy**.

    Now the application is deployed to the Elastic Beanstalk environment. You can access the application by using the URL that is displayed in the Elastic Beanstalk console.

AWS Elastic Beanstalk is undoubtedly an excellent choice for developers who want to focus on building applications and services rather than managing infrastructure. AWS Elastic Beanstalk provides a simple, scalable, and flexible platform that can help developers quickly and easily deploy applications on the AWS cloud platform.

Summary

In this chapter, we explored the world of containerization and deployment. We began by discussing what containerization is and why it is useful for modern software development. We then introduced Docker, the most popular containerization technology, and learned how to use it to package and deploy React and Flask applications.

Next, we explored the use of Docker Compose, a tool for defining and running multi-container Docker applications. We learned how to use Docker Compose to orchestrate the deployment of our applications across multiple containers.

We also delved into AWS ECR, a fully managed container registry service that allows developers to store, manage, and deploy Docker container images securely and reliably. Finally, we looked at AWS Elastic Beanstalk, a service that simplifies the process of deploying, managing, and scaling web applications. We learned how to deploy our dockerized React and Flask applications to Elastic Beanstalk with all the features with security and scalability.

In a nutshell, containerization and deployment are critical components of modern software development, and tools such as Docker and AWS services such as Elastic Container Registry and Elastic Beanstalk are essential for managing and scaling container-based applications.

We extend our sincere gratitude to you for selecting this book as your guide in mastering the art of full stack development. Your choice reflects your determination to embark on a transformative journey that combines the power of modern web technologies. It is our honor to accompany you on this path of discovery and learning.

Throughout the pages of this book, we have meticulously crafted a comprehensive roadmap to equip you with the skill set needed to conquer the realms of both frontend and backend development. We delved into the depths of React, unraveling its component-based architecture, state management, and dynamic user interfaces. Simultaneously, we navigated the intricacies of Flask, empowering you to construct robust APIs, manage databases, and handle server-side operations with finesse.

As you turn the final pages of this book, please take a moment to appreciate the knowledge you've gained and the skills you've honed. You now possess the tools to craft stunning user interfaces, harness the power of server-side applications, and seamlessly connect frontend and backend functionalities. Your journey as a full stack developer has begun, and the possibilities are limitless.

But wait, your expedition doesn't end here! As you close this chapter, new horizons await you. The world of technology is ever-evolving, and your dedication to mastering full stack development aligns perfectly with the opportunities that lie ahead. Whether you choose to build intricate web applications, design intuitive user experiences, or contribute to innovative projects, your expertise will be a cornerstone of success.

So, with your newfound proficiency in React and Flask, what's next? Perhaps you'll explore advanced React frameworks such as Next.js, dive deeper into microservices with Flask, or even embark on creating your own groundbreaking applications. The road ahead is paved with endless prospects, and your ability to shape digital experiences has never been more significant.

Once again, thank you for choosing *Full Stack with Flask and React* as your guide. Your commitment to learning and growth is inspiring, and we eagerly anticipate the remarkable contributions you will make to the ever-evolving world of technology.

Index

O

P

R

www.packtpub.com

Subscribe to our online digital library for full access to over 7,000 books and videos, as well as industry leading tools to help you plan your personal development and advance your career. For more information, please visit our website.

Why subscribe?

- Spend less time learning and more time coding with practical eBooks and Videos from over 4,000 industry professionals

- Improve your learning with Skill Plans built especially for you

- Get a free eBook or video every month

- Fully searchable for easy access to vital information

- Copy and paste, print, and bookmark content

Did you know that Packt offers eBook versions of every book published, with PDF and ePub files available? You can upgrade to the eBook version at packtpub.com and as a print book customer, you are entitled to a discount on the eBook copy. Get in touch with us at customercare@packtpub.com for more details.

At www.packtpub.com, you can also read a collection of free technical articles, sign up for a range of free newsletters, and receive exclusive discounts and offers on Packt books and eBooks.

Other Books You May Enjoy

If you enjoyed this book, you may be interested in these other books by Packt:

Flask Framework Cookbook - Third Edition

Shalabh Aggarwal

ISBN: 978-1-80461-110-4

- Explore advanced templating and data modeling techniques
- Discover effective debugging, logging, and error-handling techniques in Flask
- Work with different types of databases, including RDBMS and NoSQL
- Integrate Flask with different technologies such as Redis, Sentry, and Datadog
- Deploy and package Flask applications with Docker and Kubernetes
- Integrate GPT with your Flask application to build future-ready platforms
- Implement continuous integration and continuous deployment (CI/CD) to ensure efficient and consistent updates to your Flask web applications

Full Stack FastAPI, React, and MongoDB

Marko Aleksendrić

ISBN: 978-1-80323-182-2

- Discover the flexibility of the FARM stack
- Implement complete JWT authentication with FastAPI
- Explore the various Python drivers for MongoDB
- Discover the problems that React libraries solve
- Build simple and medium web applications with the FARM stack
- Dive into server-side rendering with Next.js
- Deploy your app with Heroku, Vercel, Ubuntu Server and Netlify
- Understand how to deploy and cache a FastAPI backend

Packt is searching for authors like you

If you're interested in becoming an author for Packt, please visit `authors.packtpub.com` and apply today. We have worked with thousands of developers and tech professionals, just like you, to help them share their insight with the global tech community. You can make a general application, apply for a specific hot topic that we are recruiting an author for, or submit your own idea.

Hi!

I am Olatunde Adedeji, author of *Full- Stack Flask and React* , I sincerely hope that your journey through the book was both enlightening and empowering, equipping you with valuable insights to elevate your proficiency in React, Flask, and the dynamic realm of full stack web application development.

It would really help us (and other potential readers!) if you could leave a review on Amazon sharing your experience with the book and how the book has expanded your skillset, introduced you to innovative techniques, or contributed to your personal journey in web application development.

Go to the link below or scan the QR code to leave your review:

`https://packt.link/r/1803248440`

Your review will help us to understand what's worked well in this book, and what could be improved upon for future editions, so it really is appreciated.

Best wishes,

Download a free PDF copy of this book

Thanks for purchasing this book!

Do you like to read on the go but are unable to carry your print books everywhere? Is your eBook purchase not compatible with the device of your choice?

Don't worry, now with every Packt book you get a DRM-free PDF version of that book at no cost.

Read anywhere, any place, on any device. Search, copy, and paste code from your favorite technical books directly into your application.

The perks don't stop there, you can get exclusive access to discounts, newsletters, and great free content in your inbox daily

Follow these simple steps to get the benefits:

1. Scan the QR code or visit the link below

https://packt.link/free-ebook/9781803248448

2. Submit your proof of purchase
3. That's it! We'll send your free PDF and other benefits to your email directly